Data and Knowledge in a Changing World

Springer

Berlin
Heidelberg
New York
Barcelona
Budapest
Hong Kong
London
Milan
Paris
Santa Clara
Singapore
Tokyo

J.-E. Dubois · N. Gershon (Eds.)

Modeling Complex Data for Creating Information

With 84 Figures and 12 Tables

Springer

CODATA Secretariat
Phyllis Glaeser, Exec. Director
51, Boulevard de Montmorency
F - 75016 Paris

Editors:

Professor Dr. Jacques-Emile Dubois
Université Denis Diderot Paris VII
Institut de Topologie et de
Dynamique des Systèmes
(ITODYS)
1 rue Guy de la Brosse
F - 75005 Paris

Dr. Nahum Gershon
The MITRE Corporation
7525 Colshire Drive
McLean, VA 22102
USA

The image on the front cover comes from an animation which shows worldwide Internet traffic. The color and height of the arcs between the countries encode the data-packet counts and destinations, while the "skyscraper" glyphs (or icons) encode total traffic volume at any site. This image was generated by Stephen G. Eick at the AT&T Bell Laboratories.

ISBN-13: 978-3-642-80201-0 e-ISBN-13: 978-3-642-80199-0
DOI: 10.1007/978-3-642-80199-0

Cataloging-in-publication Data applied for

Die Deutsche Bibliothek – CIP-Einheitsaufnahme
Modeling complex data for creating information; with 12 tables /
J. E. Dubois; N. Gershon (ed.) - Berlin; Heidelberg; New York; Barcelona; Budapest; Hong Kong; London; Milan; Paris; Santa Clara; Singapore; Tokyo: Springer, 1996
(Data and knowledge in a changing world)
ISBN-13: 978-3-642-80201-0
NE: Dubois, Jacques-Emile [Hrsg.]

Softcover reprint of the hardcover 1st edition 1996

Typesetting: Camera-ready by editors
SPIN: 10526058 51/3020-5 4 3 2 1 0 - Printed on acid-free paper

INTRODUCTION TO THE SERIES

E. Fluck
Chairman of the Publication Advisory Board

CODATA's primary purpose in launching the Series "Data and Knowledge in a Changing World" is to collect from widely differing fields the wealth of information pertaining to the intelligent exploitation of data in the conduct of science and technology and to make that information available to a multidisciplinary community. This series, in support of that goal, will provide a forum for many types of submission, both theoretical and applied. Directories and Glossaries of compiled scientific and technical data or of Space and Time Observation data could be included in a separate **CODATA** Reference Data Series.

The present series will be wide in scope as it encompasses a broad range of data and knowledge contributions, including computer-related handling and visualization of data, to the major scientific, technical and medical fields.

To this end, the series on Data and Knowledge is open to contributions of various kinds, namely those:

- Fostering the improvement, not only of the quality and accessibility of quantitative and qualitative data, but of the classical and ground-breaking methods by which numeric and symbolic data are acquired, analyzed and managed;
- presenting new data and knowledge interfaces designed to optimize interoperability and thereby increase the potential of sharing data among databases and networks;
- intensifying international cooperation in communication and data sharing-this implies works dealing with standardization, data quality agreements and conceptual data descriptions (metadata, syntactic and semantic approaches)

along with papers dealing with the evolution of Internet-based facilities, other forms of worldwide communications and electronic publishing;

- providing new insights into, or interpretations of, processes leading to creative design in the field of concurrent and/or cooperative engineering, including cognitive aspects critical to data-based decision making.

In view of the emergence of the cyclic nature of information needs in the study of complex systems, **CODATA**'s scope now includes data activities and modeling of an interdisciplinary nature such as that encountered in far-reaching projects (e.g. Global Change, various Genome projects, environmental and biodiversity issues, etc.) or in certain medical information systems. Most of the above will therefore be highlighted in this Series.

Attempting, however, to predict the contents of such series is itself paradoxical, reflecting the particular challenge in the field of data and information management whose essence is to proceed at such pace as to outstrip the very information highways it is building.

In an evolving information world in which our traditional ways of transferring information as an essential resource are being altered, this Series should capture emerging and innovative concepts thereby leading ultimately to new information paradigms. Its ambition is to be concurrently a catalyst of change and a thought-provoking forum.

ACKNOWLEDGMENT

This book is a follow up to our contribution **"The Information Revolution : Impact on Science and Technology"** in this series on Data and Knowledge. It too was inspired by the Symposium on Communication and Computer Aided Systems held at the **14th International CODATA Conference** in September 1994 in Chambéry, France.

Professor ABIR, then President of CODATA, and an international Program Committee recommended setting up some specialized panels and working sessions. In these experts came together to discuss the many crucial issues involved in the theories, methods and tools used to produce ST knowledge induced from information and data.

International Scientific Program Committee
Prof. J.-E. Dubois, France (Chairman)
Dr. F. Bisby, U.K.
Dr. M. Chinnery, U.S.A.
Prof. S. Iwata, Japan
Prof. E. Fluck, Germany
Academician F.A. Kuznetsov, Russia
Dr. J.R. Rodgers, Canada
Dr. J. Thieman, U.S.A.
Prof. A. Tsugita, Japan
Prof. Z. Xu, China

Their diverse backgrounds and experience combined for a fairly global view of the many aspects to be considered in modelizing or designing in a CAD space. It was a unique occasion to review numerous concepts both critically and pedagogically. The following sessions chaired by eminent experts produced most of the contributions of this book. Others were sollicited subsequently to complete the picture and ensure coherence

- **Knowledge and Data Interplay**

  Christophe Roche

- **Integrated Knowledge Based Systems : Uncertainty Management**

  Nahum Gershon

- **Recognition of Rigid and Flexible Objects**

  Paul Mezey

The organizing committee of the 14th International ***CODATA*** *conference is privileged to acknowledge the sponsorship of the French Academy of Sciences for the 1994 Lavoisier Celebration Year, of the International Council of Scientific Unions, the European Centre of Reflection and Studies in Thermodynamics, the University of Savoie and several organizations : ADME, SML, IRPL, AFNOR, LNE.*

Sincere thanks are expressed for the patronage of the French Ministry of the Environment, the Ministry of Higher Education and Reseach, the Ministry of Defense and the Ministry of Industry.

CODATA *is sincerely grateful for the many experts whose rich experience and reflection and hard work produced the positive and constructive discussions that took place and through which we hope to contribute to building an ever more valid information universe, since we see this as the indispensable road helping human beings to a better and more global understanding and management of the world in which we live.*

CONTENTS

Chapter 2 : Imprecision and Uncertainty in Information Systems

Chapter 3 : System Architecture and Evaluation Criteria

AUTHORS

Ana Paula AMBROSIO
Laboratoire PRISM, 45 av. des États Unis, 78000 Versailles, France
E-mail: ambrosio@prism.uvsq.fr

Eric ANGELINI
Florasynth France, 45 Bd. Marcel Pagnol, 06332 Grasse Cedex, France

Alfont BAIKER
Department of Chemical Engineering and Industrial Chemistry, Swiss Federal Institute of Technology, ETH Zentrum, 8092 Zürich, Switzerland

Jessica BARROS de SA
Laboratoire PRISM, Université de Versailles, 45 av. des Etats Unis, 78000 Versailles, France, E-mail: Jessica.Barrosdesa@prism.uvsq.fr

Jean Paul BARTHES
CNRS URA 817 HEUDIASYC, Université de Technologie de Compiègne BP 649, 60206 Compiègne, France

Stéphanie BENOIST
Société de Bio-Informatique et de Bio-technologie, 45 rue Emile Zola, 37000 Tours, France.

Jean-François BERGAMINI
Société de Bio-Informatique et de Bio-technologie, 45 rue Emile Zola, 37000 Tours, France

Hélène BESTOUGEFF
Université Paris VII- Denis Diderot, 2 place Jussieu ,75251 Paris Cedex 05, France, E-mail: bestoug@ladl.jussieu.fr

Robert BYWATER
Biostructure Department, Novo Nordisk A/S, Novo Alle, 2880 Bagsvaerd, Denmark

Daniel CABROL BASS
LARTIC, Université de Nice Sophia-Antipolis, 06108 Nice Cedex , France

Claude CACHET
LARTIC, Université de Nice Sophia-Antipolis, 06108 Nice Cedex , France

Ghislaine CARIMALO
CNAM, 2 rue Léo Delibes, 37200 Tours, France

Christophe CLEVA
LARTIC, Université de Nice Sophia-Antipolis, 06108 Nice Cedex , France

E. Richard COHEN
Science Center Rockwell International Corporation, 1049 Camino dos Rios, Thousand Oaks, CA 91360, USA

Aliette COSSE BARBI
ITODYS, Université Paris VII-Denis Diderot, 1 rue Guy de la Brosse, 75005 Paris, France

Jean Loup DELCROIX
Université de Paris Sud, Centre d'Orsay Laboratoire de Physique des Gaz et des Plasmas, Centre de Données GAPHYOR, Paris, France

Jean Pierre DOUCET
Institut de Topologie et de Dynamique des Systèmes, associé au CNRS URA 34, Université Paris VII-Denis Diderot, 1 rue Guy de la Brosse, 75005 Paris, France

Didier DUBOIS
Institut de Recherche en Informatique de Toulouse, Université Paul Sabatier, 118 route de Narbonne, 31062 Toulouse Cedex, France

JacquesEmile DUBOIS
ITODYS, Université Paris VII-Denis Diderot, 1 rue Guy de la Brosse, 75005 Paris, France

Laure DZIUBA
Florasynth France, 45 Bd Marcel Pagnol, 06332 Grasse Cedex, France

Ahmad EGHBALDAR
LARTIC, Université de Nice Sophia-Antipolis, 06108 Nice Cedex , France

Roland FELLOUS
Laboratoire de Chimie des Arômes, Université de Nice Sophia-Antipolis, 06108 Nice Cedex, France

Thomas P. FORREST
Department of Chemistry, Dalhousie University, Halifax, Nova Scotia, B3H 4J3 Canada

Claude GAUDEAU
CNRS Laboratoire de Bio-Informatique, 32 rue Emile Zola, 37000 Tours, France

Philippe GAVINI

Laboratoire de Chimie des Arômes, Université de Nice Sophia-Antipolis, 06108 Nice Cedex, France

Benoit GERARD
C.R.P.G., B.P. 20, 54501 Vandoeuvre-lès-Nancy Cedex, France

Laurent GOUTHIERE
Société de Bio-Informatique et de Bio-technologie, 45 rue Emile Zola, 37000 Tours, France

Ramzi GUETARI
LLP/CESALP, BP. 806, 74016 Annecy Cedex, France

Huw JONES
Reader of Computer Graphics, Centre for Electronic Arts, Middlesex University, Cat Hill, Barnet, Hertfordshire EN4 8HT, UK, E-mail: huwl@mdx.ac.uk

Jaap A. KAANDORP
Parallel Scientific Computing and Simulation Group, Faculty of Mathematics and Computer Science, University of Amsterdam, Kruislaan 403, 1098 SJ Amsterdam, The Netherlands

J. G. KAUFMAN
The Aluminum Association, 900 19th street, N.W., Washington D.C., 20006 2168, USA

Zbigniew KIERZKOWSKI
Technical University of Poznan, Pl. Sklodowskiej Curie 5, 60-965 Poznan, Poland

Thierry LAIDBOEUR
LARTIC, Université de Nice Sophia-Antipolis, 06108 Nice Cedex , France

Richard LAVERY
Laboratoire de Biochimie Théorique, CNRS URA 77, Institut de Biologie Physico Chimique, 13, rue Pierre et Marie Curie, Paris 75005, France

Christian. Le CARLIER de VESLUD
C.R.P.G., B.P. 20, 54501 Vandoeuvre-lès-Nancy Cedex, France

Michel LOUKIANOFF
ITODYS, Université Paris VII-Denis Diderot, 1 rue Guy de la Brosse, 75005 Paris, France

Nicolaas MARS
Knowledge-based Systems Group, Department of Computer Science, University of Twente, P.O. Box 217, 7500 AE Enschede, The Netherlands

Gilles MATHIEU
Institut de Topologie et de Dynamique des Systèmes, associé au CNRS URA 34, Université Paris VII-Denis Diderot, 1 rue Guy de la Brosse, 75005 Paris, France

Paul G. MEZEY
Molecular Modeling Group, Mathematical Chemistry Research Unit, Dept. Chemistry and Dept. Mathematics, University of Saskatchewan, 110 Science Place., Saskatoon S7N 5C9, Canada

Bruno MINDER
Department of Chemical Engineering and Industrial Chemistry, Swiss Federal Institute of Technology, ETH Zentrum, 8092 Zürich, Switzerland

Ramaswamy NILAKANTAN
American Cynamide Company, Lederle Laboratories, 401 N. Middletown Rd., Pearl River, N.Y. 10965 USA

Annick PANAYE
Institut de Topologie et de Dynamique des Systèmes, associé au CNRS URA 34, Université Paris VII-Denis Diderot, 1 rue Guy de la Brosse, 75005 Paris, France

Patrick. PIRAS
ENSSPICAM, Av. Escadrille Normandie Niemen, 13397 Marseille Cedex, France

Henri PRADE
Institut de Recherche en Informatique de Toulouse, Université Paul Sabatier, 118 route de Narbonne, 31062 Toulouse Cedex, France

Vero RAKOTOMALALA
Société de Bio-Informatique et de Bio-technologie, 45 rue Emile Zola, 37000 Tours, France

Eric RAVAUD
CNAM, 2 rue Léo Delibes, 37200 Tours, France

Michel RIVEILL
Université de Savoie LGIS, Campus scientifique, 73370 Le Bourget du Lac, France
E-mail : Michel.Riveill@univ-savoie.fr

Christophe ROCHE
LGIS, Université de Savoie, 73376 Le Bourget du Lac Cedex, France
E-mail : roche@univ-savoie.fr

Christian ROUSSEL
ENSSPICAM, Av. Escadrille Normandie Niemen, 13397 Marseille Cedex,France

Jean-Jacques ROYER
C.R.P.G., B.P. 20, 54501 Vandoeuvre-lès-Nancy Cedex, France

Olivier SCHWALM
Department of Physical Chemistry, University of Geneva, 30 Quai Ernest Ansermet, 1211 Geneva 4, Switzerland

Weiming SHEN
CNRS URA 817 HEUDIASYC, Université de Technologie de Compiègne, BP 649, 60206 Compiègne, France

Arben SHTUKA
C.R.P.G., B.P. 20, 54501 Vandoeuvre-lès-Nancy Cedex, France

Philippe SMETS
IRIDIA, Université Libre de Bruxelles, 50 avenue Franklin D. Roosevelt, CP 194/6, 1050 Bruxelles, Belgium

Gérard VANDOME
Bull/Open Systems and Software 2, rue de Vignate, 38610 Gières, France
E-mail: G.Vandome@frgu.bull.fr

Rengachari VENKATARAGHAVAN
American Cynamide Company, Lederle Laboratories, 401 N. Middletown Rd., Pearl River, N.Y. 10965 USA

Jacques WEBER
Department of Physical Chemistry, University of Geneva, 30 quai Ernest Ansermet, 1211 Geneva 4, Switzerland

Jack. H. WESTBROOK
Brookline Technologies, 5 Brookline Road, Ballston Spa , N.Y. 12020 , USA

PREFACE

J.-E DUBOIS and N. GERSHON

As with Volume 1 in this series, this book was inspired by the Symposium on "Communications and Computer Aided Systems" held at the **14th International CODATA Conference** in September 1994 in Chambéry, France. This book was conceived and influenced by the discussions at the Symposium and most of the contributions were written following the Conference. Whereas the first volume dealt with the numerous challenges facing the information revolution, especially its communication aspects, this one provides an insight into the recent tools provided by computer science for handling the complex aspects of scientific and technological data.

This volume, "Modeling Complex Data for Creating Information," is concerned with real and virtual objects often involved with data handling processes encountered frequently in modeling physical phenomena and systems behavior.

Topics concerning modeling complex data for creating information include:

- Object oriented approach for structuring data and knowledge
- Imprecision and uncertainty in information systems
- Fractal modeling and shape and surface processing
- Symmetry applications for molecular data

The choice of these topics reflects recent developments in information systems technologies. One example is object oriented technology. Recently, research, development and applications have been using object-oriented modeling for computer handling of data and data management. Object oriented technology offers increasingly easy-to-use software applications and operating systems. As a result, science and technology research and applications can now provide more flexible and effective services.

Another example of recent development is the ability to handle imprecision and uncertainty in information systems. These systems can accommodate queries that in many scientific and technological disciplines are not precisely defined. These concepts are important for both science and technology where the norm is statistically defined and where fluctuations from the norm are expected. Chapter 2 deals with such systems.

A third example is the ability to handle, manipulate, and visualize physical 3D objects in real time. 3D representation through both classical and fractal methods is essential for our visual approach to reality. It is also a powerful concept for handling various overlapping data and information that can be mapped in a particular 3D space. A few papers in this volume tackle symmetry problems and fractal modeling.

Concepts and theories are still among the essential and basic challenges facing our Information Society. However, any knowledge emerging from raw data often involves the handling of complex data and the elaboration of new interpretations. This book discusses such processes in a multitude of fields with an eye to creating valid and useful new information.

Acknowledgments: We would like to thank all the authors of this book for their contributions and hard work. We would also like to acknowledge the hard work and dedication of Ms. Bernice Dubois in helping to bring this book to the world and the efforts of Dr. Belhadri Messabih in putting together the contributions of authors from so many countries.

PART ONE

INFORMATION AND DATA INTERPLAY

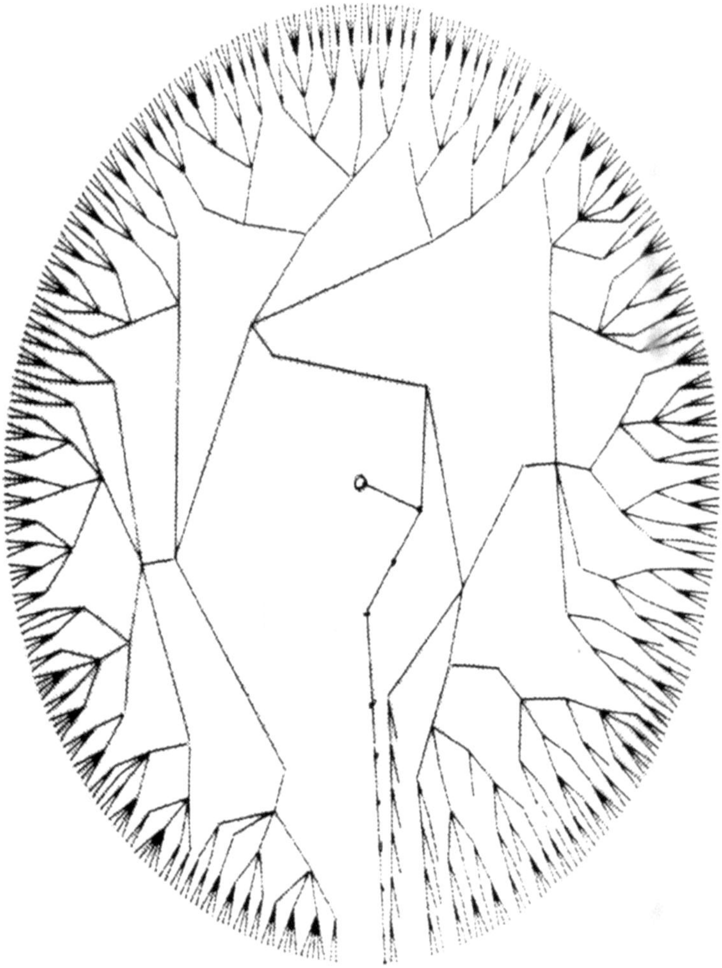

DARC Artificial Intelligence Organization of a complex family of C_nH_{2n+2} (1≤n≤11) compounds in an ordered graph (hyperstructure HS). Such a space is useful to establish property-relationships dealing with the internal atomic topology of each compound node and the 2D or 3D vicinity of the disconnected topological nodes of this HS. (J.-E. Dubois "Chemical Application of Graph Theory", A.T. Balaban, Academic press 1976 p.333)

Chapter 1

STRUCTURING DATA FOR INTEGRATED KBS

How to Structure Data: the Object Oriented Approach (a Survey)

Christophe ROCHE

L.G.I.S., Université de Savoie 73376 Le Bourget du Lac Cedex, France
E-mail : roche@univ-savoie.fr

Abstract

The complexity of some problems increasingly requires explicit representation of knowledge, as well as of its links and processing. This kind of approach corresponds to a desire to formalize and organize data in order to better control them, and to move from data to knowledge. The purpose of this article is to review the problems raised by this kind of effort. In particular we will look at the problem of choosing a suitable representation formalism, especially object-oriented representations, as well as at the problems of storage in data bases and of sharing those knowledge bases.

Resume

La complexité de certains problèmes nécessite de plus en plus une représentation explicite des connaissances, de leurs liens et de leurs traitements. Une telle démarche correspond à une volonté de formaliser et d'organiser les données afin de mieux les maîtriser, de passer des données aux connaissances. Cet article a pour objectif de passer en revue les problèmes que soulève une telle démarche. Nous verrons en particulier le problème du choix d'un formalisme de représentation adéquate, et plus particulièrement les représentations orientées vers l'objet, ainsi que le problème du stockage dans des bases de données et celui de la répartition de ces bases de connaissance .

1 INTRODUCTION

Processing a large number of data is a familiar problem for which satisfactory solutions exist. This is especially the case with methods of numerical and statistical analysis. However, conventional approaches rely on rigid programs and data models detrimental to the comprehension, maintenance, re-use and extension of those systems. In fact, those approaches do not permit data to be structured and organized, nor any representation of real-world entities and their relationships. Furthermore, the complexity of some problems requires a qualitative approach rather than a quantitative one, based on explicit representation of knowledge, links between pieces of knowledge, and knowledge processing. This approach is meant to formalize and organize data in order to better control them, and to move from data to **knowledge.**

The central problem that then arises is to choose a formalism to represent knowledge, whether data, concepts, relations or treatments. This problem is compounded by many other problem sets, in particular those of knowledge acquisition, storage in appropriate data bases and sharing of those knowledge bases. It is not possible within a single article to answer all these questions. We will focus on the central problem of the choice of a knowledge representation formalism, with the concept of the object - today an unavoidable one - as it results both from knowledge representation and from object-oriented programming. Then we will present the two major problems of storage in object-oriented databases and of sharing knowledge bases.

2 OBJECT-ORIENTED KNOWLEDGE REPRESENTATION (FRAME FORMALISM)

Frame formalism is certainly the most commonly used in Descriptive Knowledge Representation. It results from the work that has been done both in Knowledge Representation and in Artificial Intelligence or Psychology. The first computer application of these models dates from 1975 [1]. Since then many systems have been developed, from FRL [2] and KRL [3] at the end of the 1970s, to systems including many functionalities such as *SRL* [4] or *Unit* that we will see at the heart of expert system development environments such as *Knowledge Craft* [5] or *Knowledge Engineering Environment* [6]. For a general presentation of those systems, the reader may refer to [7]. Frame-based representations have two goals, to model entities from the real world and conceptual abstractions that can be defined by a finite number of properties and express relationships between those notions. This is why frame formalism relies on two notions, the first being the inseparable triplet of "frame-slot-facets" and the second that of relationship.

2.1 Frame-slot-facets

The object, a frame, is a structure defined by a set of « attribute-value » couples that is inseparable from the frame. These attributes, also called properties or slots, are directly accessible. The definition of these attributes is complemented using facets. Thus there are declarative facets that concern the value of an attribute: type, value domain, default value. There are also procedural facets to permit data driven programming. These procedures are activated each time the value of an attribute is accessed : in reading, writing, deletion or absence of any value.

2.2 Relations

The second fundamental concept in these systems is that of the relation. In fact, real-world entities and their abstractions maintain many different semantic relations. We could mention as an example the relation of « specialization-generalization » that organizes concepts in the form of a hierarchical graph. Of course there are other relations, some general, such as relations of « part-of », « composed-of », « uses », and others specific to a given application. The semantics of these last relations then depends on the application field and can be defined by the programmer.

3 OBJECT-ORIENTED LANGUAGES

Object-Oriented Languages stem from classic software engineering problems. These problems are usually met when writing and using large software : specification, design, maintenance and reusability. These languages provide a really new programming approach which is very different from classic programming where we encounter two different types of entities that are physically distinct although there is no semantic separation between them: data and programs. This is not the case with object-oriented programming. There we have only one type of entity, the object, which associates in a single entity both the data and the procedures for its use. The object is then a re-usable module, that can be exploited through an interface that also protects the data.

These programming languages, resulting from Simula 67 [8] and made popular by Smalltalk 80 [9], are based on four concepts. The concept of object was defined above. The concept of class can be seen from different viewpoints: as a type of object or as a set of objects with similar descriptions and behaviors. In the concept of message the only way to manipulate objects is to send them messages to execute the procedures associated with the object. Finally, the class hierarchy concept makes it possible to structure and share object programs by generalizing and specializing existing classes.

4 OBJECT-ORIENTED DATA BASES

Object-oriented data base management systems represent a field of interest that touches on conventional data bases and object-oriented programming [10], [11]. In fact, object-oriented programming is concerned, through its evolution, with concepts from conventional data bases applicable to objects : perennity, storage, concurrence, restoration after a breakdown, etc. At the same time, the appearance of new applications such as CAD, office automation, Artificial Intelligence and image processing have resulted in new needs that conventional databases can hardly satisfy, e.g. the need to define and use entities with more complex structure and behavior. The new generation of Database Systems based on the object model allows these new needs to be met by offering different services: definition of objects with a complex structure, interface facilitated with the object-oriented languages, taking into account the dynamics (methods) of the objects during their storage. At present there is no unique object model for object-oriented databases. There are therefore as many object models, with their own vocabulary, as there are systems. The concepts used mainly involve notions stemming from object languages, including some characteristics of the representations based on frames. We then find again the notions of class, object (from a class), method (operation on an object), hierarchy between classes and procedural attachments.

For example, let us consider the object-oriented database Orion developed by the "Microelectronics and Computer Technology Corporation" (Austin, USA) in 1987. It was mainly designed for Artificial Intelligence applications, multimedia and computer-aided design. The definition of an Orion class is similar to that of an object-oriented language (Orion supports multiple inheritance, so it is possible to define a class using several existing classes to inherit their descriptions). A new object is created using a specific function, *make*, defining as parameters the name of the class and the value of its attributes. As for object manipulation, it is done in a similar fashion to traditional data bases using various functions such as: *select, delete, change*. To conclude this brief presentation of Orion, let us note that the set of definitions of classes with their hierarchical links represents the semantics of the object data base. Orion ensures the cohesion of this set during any structural modification of the classes and of their hierarchy : adding or deleting attributes, adding or deleting methods, modification of the heritage mode, adding or deleting superclasses.

Certain relatively recent firms offer purely object Database Systems, for example Ontos, Object-Design, O2 and Versant. Nevertheless, it would be rash not to take what exists into account. The majority of operational data base management systems are relational systems. For this reason the different models, object and relational, must cooperate and interface with each other. This approach has been adopted by the publishers of most relational databases who offer mixed « relational-object » approaches. For example, AskGroup proposes an object extension for its relational database Ingres; Oracle plans to offer it in its version 8 and IBM has committed itself with its Starburst project. SQL requests can then be masked with tools using more user-friendly interfaces that are close to the archetypal object language, Smalltalk80. On the other hand, object-oriented databases are trying to interface with relational databases: products offered by O2 , Object Design and Versant all allow an SQL interface. The final objective for the user is to be able to manipulate objects without being required to know the SQL language or the relational database (the correspondence between the objects and the tables is managed by a specific interface). The reader can refer to [12] for a presentation of the evolution of data-modeling for databases, and to [13] for a recent set of references in the Object-oriented data base management systems field.

5 The Development of Large, Shared Knowledge Bases

To be able to structure and organize our information to obtain major knowledge bases and to be able to save them is not sufficient. It must be possible to share the acquired knowledge. It is surprising to observe that a large number of researchers are working on the problem of knowledge representation and Knowledge-Based Systems, but that very little work can be shared. Too many researchers reinvent the ontological wheel and too much work is evaluated in toy worlds. Shared Knowledge-Bases must be public resources for research and applications. Furthermore, such shared Knowledge Bases could have great practical utility in industrial applications, particularly in engineering fields.

What then is the problem? Part of the answer is that researchers have no common language for communicating knowledge. Let us note that sharing knowledge is considerably more complicated than sharing facts. As a matter of fact, sharing knowledge requires agreement on the meaning of symbols, assuming that it is possible to share ontologies. It means sharing networks of categories or classes, and their relationships in an inheritance network. Sharing Knowledge Bases also implies being able to share inference mechanisms on them. A working proposal was created to reach these goals through a computer language called KIF, Knowledge Interchange format [14].

KIF has the objective of interchanging knowledge among different programs written by different programmers in different languages. It cannot be a common language, but it offers an **interchange** format. That means that if someone wants to use a Knowledge Base written in KIF, he has to first convert the data into his own internal form (particular frames, arrays, list and so on). On the other hand, if someone wants to share his Knowledge Base he has to translate that knowledge into KIF before delivering it as text.

The BRS system [15], or Binary Relation System, proceeds in the same way. BRS, developed at LGIS, is a programming language based on binary relations which provides many advantages. It permits the definition of new knowledge representation languages according to needs and it plays an *interlingua* role for knowledge bases like KIF. Finally, it authorizes a satisfactory normalization of object-oriented systems built using BRS so that, in addition to traditional logic, the relations theory can be used. For a general presentation of development of large and shared knowledge Bases see [16].

KEYWORDS

Knowledge representation, frame system, object-oriented language, object-oriented database, shared knowledge base.

REFERENCES

[1] Minsky M., A framework for representing knowledge, The psychology of computer vision, Winston (eds), Mc Graw-Hill (1975).
[2] Roberts B.R., Goldstein I.P., The FRL Primer, Massachusetts Institute of Technology, Artificial Intelligence Laboratory, Memo n° 408, July (1977).
[3] Bobrow D.G. , Winograd T., An overview of KRL, a Knowledge Representation Language, CSL-76-4, July 4, Xerox Palo Alto Research Center (1976).
[4] Wright J.M., Fox M., SRL /1.5 user manual, Intelligent Systems Laboratory, Carnegie-Mellon University, Draft of 1st Dec.(1983).
[5] Carnegie Group, Knowledge Craft 3.0, reference manual, Carnegie Group Inc. Pittsburgh Pennsylvania (1985).
[6] Intellicorp, KEE, software development system user's manual, version 2.1, Intellicorp (1985).
[7] Stefik M., Bobrow D., Object-Oriented Programming : Themes and Variations, AI Magazine, Vol. VI, n°4, Winter (1986).
[8] Dahl, O. Jordan "Simula, An Algol based simulation language" Communication de l'ACM 9,9 p.671-681, july (1966).
[9] Goldberg A., Robson D., Smalltalk80 : The language and its implementation, Addison-Wesley Publishing company
[10] R., Catell ,Object Data Management : Object-oriented and extended relational databases systems Addison-Wesley (1991).
[11] Kim W., Lochovsky E.,Object-oriented concepts, databases and applications, Addison Wesley (1988).
[12] Shamkant B. Navathe , Evolution of Data Modeling for Databases, Communication of the ACM, September, Vol. 35, n°9 (1992).
[13] Zdonik S., Maier D. (eds), Readings in Object-Oriented Database Systems , Morgan Kaufmann (1990).
[14] Genesereth M.R., Fikes R.E., Knowledge Interchange Format, version 3.0 reference manual, Report Logic-92-1, Logic Group, Stanford University, June (1992).
[15] Roche C., Deloule F., BRS : A generic-binary relation system for knowledge-based systems, International conference on Modelling Languages for KBS,. Amsterdam, Jan. (1995).
[16] Gruber T.R. , The development of Large, Shared Knowledge-Bases : Collaborative Activities at Stanford, Technical Report KSL 90-62, Knowledge Systems Laboratory, Stanford University, Aug.(1990).

INTELLIGENT AGENTS : A NEW STEP TOWARDS DATA AND KNOWLEDGE ORGANIZATION

Hélène BESTOUGEFF

Université Paris 7 - Denis Diderot 2, Place Jussieu 75251 Paris Cedex 05, France, E-mail : bestoug@ladl.jussieu.fr

ABSTRACT

Using structured information as knowledge bases and an object oriented approach allow the design of entities which are able to reason but not to act in an environment driven by internal goals. Moreover, at this level, it is not possible to model cooperation between entities which lead to a distributed organisation of knowledge bases.

The aim of this article is to show how the concepts of intelligent agents and multi-agent systems can provide a new way to structure data and knowledge bases in order to include various types of actions.

RESUME

La structuration des informations en bases de connaissances et l'utilisation de l'approche objet permettent de construire des entités qui ont la capacité de raisonner mais non celle d'agir dans un environnement mues par des objectifs internes. De plus, il n'est pas possible à ce niveau de modéliser les coopérations entre entités conduisant à une organisation distribuée de bases de connaissances.

Le but de cette article est de montrer que les concepts d'agents intelligents et de systèmes multi-agents sont de nouvelles facons de structurer les données et les bases de connaissances afin d''intégrer divers types d'actions.

1 INTRODUCTION

To handle complex information, the first step was to move from data to knowledge. Expert systems, or more generally knowledge based systems, were designed in order to access or create new information by reasoning. The main problems here were knowledge acquisition, knowledge representation and reasoning and management of large and shared knowledge bases [1]. The next step is to be able to distribute and then pool a common expertise. This involves not only intelligent behaviour but communication protocols between autonomous entities in order to exchange information and solve problems through cooperation.The intelligent agent approach can satisfy these new goals provided some questions on structure of individual agents and overall cooperation are answered. Below, we first present the general concept of agent. We then briefly describe some agent languages and systems. Finally, we discuss the role of agents in the organization of data and knowledge.

2 WHAT IS AN INTELLIGENT AGENT ?

As with many new concepts, no unified definition for an agent exists thus far. It is however possible to list some common desirable characteristics [2], [3], [4].

- autonomy: an autonomous agent can operate and control itself independently with respect to some *goals*. It is an active entity which does not blindly obey external commands: it can modify requests, ask for clarification or even refuse to satisfy a request,

- social ability: agents communicate with other agents via explicit language,

- reactivity: agents are open systems in that they can react to their environment in synchronic or diachronic fashion,

- pro-activeness: agents can initiate transformations independently of the environment modification.

So far, with these features, agents can be seen as specific software processes that encapsulate some state, have some knowledge with inference capability and are able to communicate with other agents.

Another, more anthropomorphic, view is proposed by different researchers whose idea is to model complex systems organization with abstractions taken from human interaction. For instance Y.Shoham[5] defines an agent as "an entity whose state is viewed as consisting of mental components such as beliefs, capabilities, choices and commitments". Others introduce emotional agents. In this way, agents can be seen as either automated or human.

In this set up some additional characteristics can be added such as:

- veracity: agents will not knowingly communicate false information,

- rationality: agents are not schizophrenic, they will act in order to satisfy their goals.

Agents have generally only a partial representation of their environment; therefore they are not omniscient.

For aplications related to electronic networks, another important attribute is *mobility*. The agent is able to migrate from machine to machine in a heterogeneous network (see below).

We can summarize all these features by saying that an agent is an "intelligent" entity whose behaviour, (actions and communications), is oriented to the satisfaction of its goals using all available resources.

Agents can act as individuals or within multi-agent systems. In this latter case, it is necessary to define inter-agent protocols to manage the coordination of agents.

From a general view point agent *interactions* can be described on three levels:

- information: this concerns facts and knowledge that must be transmitted (also called the information content),

- intention: the same information can be communicated with different intentions, e.g. ask about a fact, deny a fact, tell about a fact ...,

- communication: this concerns the unified protocols governing agent interaction

3 AGENT LANGUAGES AND SYSTEMS

With the above list of characteristics one can build quite different agent architectures. In particular, the notion of intelligent agent can be understood differently, i.e. the agent may have a predefined knowledge base and an inference mechanism, a knowledge acquisition system or, even better, some learning capacity. From a user approach one can ignore the details of the internal structure of agents but must understand the communication language defining the interface between agents, (similarly one ignore the internal data structures and algorithms of objects but use the message passing facilities).In fact three main questions must be answered:

- How agents send and receive messages ?

- What is the meaning of an individual message ?

- How agents structure their conversations ?

3.1 Languages

Different languages are based either on a *procedural* approach or on a *declarative* approach. In the first case, communication is carried out through procedural directives. Individual commands and entire programs can be transmitted. A program can choose when and where to migrate and suspend its execution at an arbitrary point. *Tcl* and *Telescript* from General Magic, Inc. belong to this type. Tcl is an embeddable scripting language that is highly portable and freely available.

Several systems are currently built using this language. In *Telescript* two main concepts are defined: place and agent. Places are virtual locations that are occupied by agents which are *mobile software processes.*

In the latter case, all exchanges take place via declarative statements. One such language is ACL developped in the context of the "DARPA Knowledge Sharing Effort" [4]. This language can be described in three parts:

- the vocabulary, which is a list of words concerning some common applications, for instance, a vocabulary describing 3D geometry (ontologies),

- a content language called KIF which is a prefix version of first order predicate calculus with different extension [6],

- a message language called KQML [7]. A KQML messsage is called a *performative*. This term is taken from speech act theory: it was shown that making some statements (speech acts) using special classes of utterances such as "I promise", "I demand",..a speaker makes a commitment to perform some action by virtue of saying it. An open list of predefined performatives are proposed in the KQML report.

An ACL expression is a KQML message where the arguments are KIF terms (or sentences) formed by words taken from ACL vocabulary.

3.2 Systems

One popular area of applications is the development of intelligent agents for the World Wide Web [8]. Such agents can perform information filtering by searching through the Web for the pertinent information using indexing techniques; they can also guide people in answering FAQs (process files of frequently asked questions), or perform more sophisticated tasks such as expertise location. Several agents have been defined for e-mail applications such as automatic message transfer, alerting users to special messages, extracting specific information from messages... Commercial products are already available. The above applications can be characterized as integrated and goal oriented interfaces.

People interested in distributed AI have worked since the mid-1980's on issues related to multiagent-systems. Cooperative distributed problem solving (CDPS) can be seen as *multi-agent* systems which cooperate to solve problems[9].The organization of communication is either direct, where agents organize their own coordination, or through specific agents sometimes called *facilitors.* This approach to software interoperation has been applied in such domains as concurrent engineering and database integration.

Finally, using additional anthropomorphic features, multiagent-systems are a new way to construct and simulate artificial worlds and social communities. Individuals can be directly represented by agents with their behaviour and interactions.The main advantage is to be able to consider at the same time quantitative and qualitative parameters and use the system as an experimental laboratory where the user can choose different individuals, modify their behaviour and change conditions of the environment.

4 ROLE OF AGENTS IN DATA AND KNOWLEDGE ORGANIZATION

Agent models offer new abstractions for representing data and their processing that extend those proposed by object models. As far as a single agent is concerned, the attitude towards information is not just a retrieval process (direct or through inference procedures); the *intention level* introduces notions such as belief, desire, obligation, commitment... It allows, on the one hand, the description of the perception that an agent has of its environment: what is believed, at what time and under what conditions, and on the other hand, the characterization of the way an agent will perform its actions in relation to other agents e.g. desire or commitment. Hence, each piece of information is not only defined in terms of a structure (syntax and semantics) but also in terms of its use (pragmatics).

If then we consider multi-agent systems, the important point is to define the organizational structure of the net: distribution of its knowledge, control and problem solving capabilities.Here, it is assumed that no single expertise (one agent) can solve the problem, some cooperation must be established. In the same way as humans try to solve a complex problem through common work, it is necessary to define the local knowledge of each agent, the possible inconsistencies between these different knowledgebases, the resolution of conflicting goals... The new problem is not so much sharing the knowledge as the dynamic construction of a pool of knowledge for a specific goal which constitutes a model of agent aggregation.

5 CONCLUSION

Today, a huge amount of data organized in data bases or knowledge bases is available with computer tools, including programs for solving standard analytical problems or sophisticated expert systems. However, most of the tools provide point solution and are based on idiosyncratic representations and algorithms. One way to face problem complexity is to define *cognitive problem solvers* in terms of intelligent agents making it possible to construct models closely related to human processes which can be seen as *integrated knowledge based systems*[10]. The intelligent agent paradigm is still in a developping stage but new tools and standards should be available soon [11].

KEYWORDS

intelligent agent, multi-agent systems, distributed problem solving, cognitive problem solver.

REFERENCES

[1] Bestougeff H., Intelligent Information Systems: from Data to Knowledge, CODATA International Workshop on Decision Making in Science and Technology, Université de Marne La Vallée - France, Nov (1993).
[2] Ferber J. Les Systèmes Multi-Agents - Vers une Intelligence Collective, InterEditions, Paris (1995).
[3] Wooldridge M., Jennings N.R., Agents Theories, Architectures and Languages - a Survey, Intelligent Agents, ECAI Workshop on Agent Theories, Architectures and Languages. Lecture Notes in Artificial Intelligence 890, Springer Verlag (1994).
[4] Communications of the ACM, Special Issue on Intelligent Agents 37 (7), July (1994).
[5] Shoam Y., Agent-Oriented Programming, Artificial Intelligence 60 (1993).
[6] Genesereth M.R., Fikes R.E, Knowledge Interchange Format, Version 3.0 Reference Manual, Technical Report, Computer Science Dept. Stanford University (1992).
[7] Finin, T., & al., Specification of the KQML Agent Communication Language, the DARPA Knowledge Sharing Initiative External Interfaces Working Group (1992).
[8] Etzioni O., Weld D.S., Intelligent Agents on the Internet, IEEE Expert, August (1995).
[9] Durfee E.H., Lesser V.R., Corkill D.D, Trends in Cooperative Distributed Problem Solving, IEEE Transactions on Knowledge and Data Engineering, Vol.1, N°1, March (1989).
[10] Bestougeff H., Cognitive Problem Solving: an Intelligent Agent Approach, Japan, International SAMPE Symposium & Exhibition, Tokyo, Sept (1995)
[11] An entry point for information on agents on INTERNET http://www.info.unicaen.fr/~serge/sma.html

DESCRIPTION AND APPLICATIONS OF AN OBJECT-ORIENTED MODEL PDM

Weiming SHEN and Jean-Paul BARTHES

CNRS URA 817 HEUDIASYC, Université de Technologie de Compiègne BP 649
60206 Compiègne, France

ABSTRACT

Most industrial applications in the domains of CAD/CAM/CASE, KBS/DBS as well as that of robotics need to handle a large volume of information whether knowledge or data. In general, these applications require the representation of complex data elements as well as complex relationships among them. Users in these environments have found traditional technology inadequate in terms of flexibility, modeling power and efficiency. Thus, it becomes necessary to develop environments capable of supporting the various paradigms used in large projects. Such environments require a good object representation, efficient programming features, adequate mechanisms for storage and concurrence control and good communications with other software systems. This paper presents the main features of an object-oriented representation model PDM (Property Driven Model) and its implementations. An example is presented for showing the application of the PDM model and the object-oriented programming environment MOSS in the field of mechanical CAD.

RESUME

La plupart des applications industrielles dans les domaines de CFAO, SGBD/BC, Robotique demandent le traitement d'un grand volume d'informations. En général, ces applications exigent un modèle de représentation pour des données complexes et des relations compliquées entre celles-ci. Il est nécessaire de développer des environnements pour supporter différents paradigmes utilisés dans les grands projets. De tels projets nécessitent, entre autre, un bon modèle de représentation orienté-objet, un langage de programmation puissant, des mécanismes de sauvegarde des données et de contrôle de cohérence et une interface de communication avec d'autres systèmes. Dans cet article, nous présentons les caractéristiques principales du modèle de représentation orienté-objet PDM (Property Driven Model) et la mise en oeuvre de celui-ci. Un exemple d'application du modèle PDM et du langage de programmation orienté-objet MOSS dans le domaine de CAO en mécanique est presenté dans cet article.

1 INTRODUCTION

Most industrial applications like CAD, KBS/DBS and robotics need to handle a large amount of information whether knowledge or data. Such information is linked by a web of complex relationships. Traditional programming environment and database technology cannot meet such requirements. At the same time, some researchers in the field of artificial intelligence advocate abandoning small examples in favor of tackling larger real world applications [Schank 91]. The CYC project [Lenat & Guha 90] is a good example. Thus, it becomes necessary to develop environments capable of supporting the various paradigms used in large projects. Such environments require: (i) a good object representation, (ii) efficient programming features, (iii) adequate mechanisms for storage and concurrence control, (iv) good communication with other software systems.

Artificial intelligence can provide very expressive models for solving the first requirement. Obviously, one turns to object-oriented programming for programming features, and to databases for storage and concurrence control. Communication and integration are found in the domain of software engineering, and they are nowadays one of the most challenging issues when using the object paradigm in complex applications.

Considering the above requirements, it seems interesting to take an approach based upon the concept of object, for two reasons: (i) the concept of object is common to artificial intelligence, programming languages, and databases; (ii) it is also a convenient mechanism used in software engineering for integrating heterogeneous pieces of software.

Many object models have been proposed both in artificial intelligence and for databases. In artificial intelligence, models tend to act as prototypes and repositories for default values. Many models have been used in knowledge representation languages like KRL [Bobrow & Winograd 77], RLL [Greiner & Lenat 80], SRL [Fox et al 86], and many others which have frame editor capabilities. Within the database community recent attempts have been made to cope with the need to support efficiently deductive databases and applications like CAD. Object models are proposed like in [Yu et al 86], in O2 [Lécluse et al 87]. From the database perspective object models tend to act as specifications for storing information.

In this paper, we present the main features and the recent developments of a specific representation model PDM (Property Driven Model) and an object-oriented programming environment MOSS (Multi-level Object Simulation System). Then we present a small example showing the application of the object-oriented representation model, PDM, and the object-oriented programming environment MOSS in the field of mechanical CAD systems.

2 AN OVERVIEW OF PDM

PDM was developed around 1977 at the University of Technology of Compiègne [Barthès et al 79], and was influenced by Minsky's frame model [Minsky 75], Abrial's semantic model [Abrial 74], and the metaphysics of Thomas Aquinas (XIIIs) who considered that the world consists of entities (objects) having quiddities (properties), and that quiddities are also objects. This approach was motivated by the early use of representation in the domain of robotics, where objects may have well defined properties—acquired through various sensors—but do not belong to any specific class until they are fully recognized. Thus, PDM can represent objects as soon as properties exist.

PDM was designed to be highly dynamic; it includes classless objects, distinguishes between specification and prototyping; it is reflective; properties are shared and can be structured as trees; object behavior is specified by three kinds of methods including universal methods attached to nothing in particular.

A PDM object is described by a set of properties, e.g.

```
has-name:                      EINSTEIN
has-first-name:                ALBERT
has-birth-country:             Germany
has-speciality:                Relativity
```

The two first properties have values directly attached to them and are called *terminal properties*; the two last ones are linked with other objects and hence are called *structural properties*. The terms stand for historical reasons and were proposed in 1977; today, following the OMG recommendations, we should call them *attributes* and *relationships*. PDM does not use the classical three-level frame-slot-facet format usually found in frame languages. Instead properties are themselves declared as objects and described by using the same format, e.g.,

```
<has-name>
   has-property-name:          NAME
   is-terminal-property-of:    PERSON
   has-method:                 =if-added,
                               =print-self,
                               =entry
```

A class PERSON is also an object, and can be described as follows:

```
<person>
   has-entity-name:                      PERSON
   has-terminal-property:                has-name
                                         has-first-name
   has-structural-property:              has-speciality
                                         has-birth-country
   has-method:                           =summary,
                                         =print-self, ...
```

In the above example it can be seen that information usually found in slots is factored in the attribute objects. Methods are also objects. Classes in PDM can be created when there are enough objects of the same kind, which is fairly common in artificial intelligence. Once a class is defined, it can be used as a generator (pattern) for creating additional objects or instances.

PDM uses a four-layer specification hierarchy, i.e., instance layer, model layer, meta-model layer and meta-meta-model layer (see Fig.1). The class PERSON is represented in the model layer which contains classes and instances of properties. The model layer is a specification layer describing the format of objects in the instance layer, as shown by the dashed lines in Fig.1. Of course, properties themselves and classes have their own specifications, described in a meta-model layer. In turn, the objects on the meta-model layer have a general format described in a meta-meta-model layer which contains the object ENTITY. As usual in this kind of approach, ENTITY is its own specification, i.e., there is no meta-meta-meta-model layer. Thus, a *structural reflectivity* is achieved at this level. The two higher levels define the internal structure of the chosen representation, and constitute the kernel of the system. *Reflectivity* is a powerful concept, since it allows to modify dynamically the structure of the model itself; however it is difficult to handle. *Reflectivity* is an extremely useful feature for automatically providing explanations, or for implementing learning mechanisms.

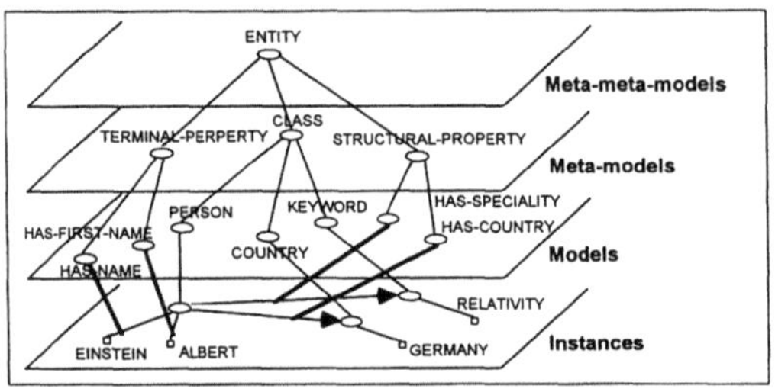

Figure 1. A four-layer specification hierarchy

Like most other object representation models, PDM allows a super-class/class/sub-class (meta-models/models/instances) hierarchy through IS-A structural property.

Inheritance is defined within each layer of the specification hierarchy. An interesting feature of PDM is the possibility of defining IS-A links between instances, thus introducing a prototyping mechanism. This can be used in particular at property level for particularizing a more general property.

Another interesting feature is to create classless objects. The concept of classless objects is useful for recognition processes, e.g., scene analysis in robotics.

Contrary to what is found in most models, in PDM, values associated with terminal properties are not restricted to any specific type, and can be anything from a string to an image or an executable function (which is useful for implementing demons). However they can be restricted by using the demons attached to the specific terminal property thus implementing data types. Such demons are invoked at creation or modification time or for displaying values.

Terminal properties may be inverted, creating external references to objects called *entry-points*. Such inversions occur at creation time, and are performed by an appropriate demon. Entry-points are automatically maintained and updated when values are modified. Entry-points are not required to be unique, which simplifies indexing.

Comparison operators may be user-defined and associated with terminal properties. Structural properties are automatically inverted, which allows one to navigate easily within the database, and avoids extra indexing structures.

Each entity (all objects, classes and instances) has an internal unique identifier, its internal key. Any object can grow or shrink at any time. Classes are modeled as objects providing an integrated data dictionary. Objects are instantiated by using the classes in a similar way as in SMALLTALK. PDM is a consistent object system, i.e., every object has a unique identifier, and there are no dangling references.

The PDM representation is based on the concept of property (attributes and relationships). Properties exist independently of any class. Once a property exists, it can be used to characterize objects. Consequently, an object can be interpreted independently of any class, as long as the corresponding properties it contains exist in the system.

3 IMPLEMENTATION OF THE PDM MODEL

Choosing a data structure for implementing the PDM representation involved several constraints: (i) the dynamic aspect of the objects; (ii) the capacity for a large number of objects; (iii) the problems of object sharing. The environment that resulted from the implementation of the PDM model is called MOSS [Barthès 89].

The LISP environment provided the dynamic features with the use of lists rather than "defstruct". Simple association lists were used in MOSS for allowing storage of an object as a sequence of <property, list of values> pairs. Specifying the property explicitly in a LISP environment is virtually free. Traditional frame languages on the other hand express an object as a 3-level structure frame/slot/facet, where facets are sometimes used to contain meta information. Names for facets are often predefined, and characterize the resulting frame representation language. With the PDM approach, properties—corresponding to slot names—are first class objects. Thus, the property name refers to a separate object, and facets are in turn expressed as properties of that object. Hence, facets are factorized for all objects sharing the property. This is very clean and efficient in terms of storage, which is a good point for handling a large number of objects. However it has the major disadvantage of not allowing addition of specific (meta-)information to qualify the property of a particular object.

The overall MOSS architecture is shown in Figure 2.

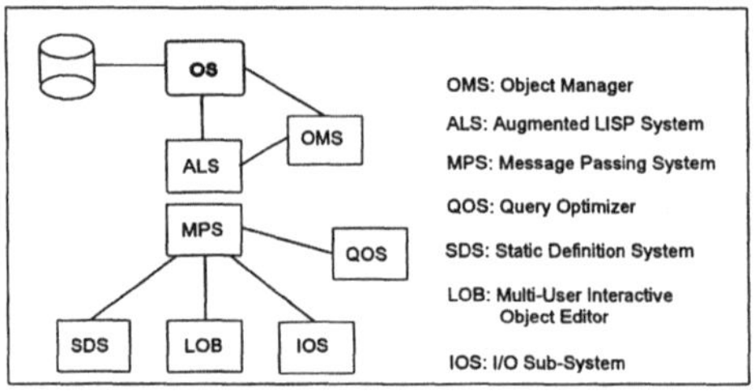

Figure 2. Overall architecture of the MOSS system

In recent years, with the progress of research, several important features have been added to MOSS.

- *Persistency*

Since we have to handle a large number of objects, we need to keep objects in secondary storage and load them into working store only when necessary [Barthès et al 90]. However LISP environments are well known for their poor access methods (streams) obliging us to resort to back end databases for storing data. Such a solution is not well adapted to a very large number of objects.

In addition conventional relational databases show a poor behavior when one wants to add dynamically new attributes, or to create new individual links among objects.

We adopted the strategy of storing objects in a flat format contiguously on disk. The LISP environment is used as a cache for objects during transactions. A LISP/disk interface MGF [Barthès 85]was designed to allow one to read and write a single object from or onto disk. Special disk and in-core formats were investigated so as to minimize reformatting time.

- *Versioning*

The versioning mechanism is very useful in the domain of CAD. Versioning can be used for exploring parallel hypotheses, or for doing concurrent design, letting design teams diverge, then reconciliating the separate designs [Barthès 93].

Each object contains a set of attributes with associated values. A new version is created when a new set of values is specified for an attribute. MOSS stores the changes within the object itself, associated with a transaction stamp. This is possible due to the dynamic format selected for implementing the objects. In parallel, the relationship between stamps is stored in a version graph which records the possible states of the database, thus providing a viewpoint mechanism similar to that of ART [Clayton 85] and of O2 [Lécluse et al 87].

The global version graph is not a tree, since recombination of versions can lead to merged branches. In the case of merged versions, there is no ambiguity unlike the case for multiple inheritance in semantic models.

In the MOSS system all objects are versioned, including classes and methods. However, although consistency is enforced between classes and instances, there is no provision for a mechanism associating new classes with old instances, or new instances with old versions of classes.

In the MOSS system, the end user has no control over the version mechanism. A new version is automatically created at each new general update. A new revision can be created by collapsing all information onto a single node. The other nodes can then be archived. The process is somewhat similar to the "believe" command in ART.

- *Concurrence*

Concurrence is used to allow several users to work simultaneously on the same information. A good discussion can be found in [Kim et al 87]. A traditional approach for solving concurrence problems has been to use locking. In MOSS, we took a different approach.

MOSS uses an optimistic concurrence approach, i.e., when an update is performed; all operations are checked for consistency with the previous operations . If none of the integrity constraints is violated, then the version is added linearly to the version-graph. If some inconsistency is found, then the user is notified and may elect either to modify the data to comply with the recorded modifications, or to create an alternative database version.

MOSS was first developed in an augmented VAX/VMS LISP (UTC-LISP) environment around 1987. At present, MOSS has four versions: Version 3.1 on UNIX (SUN Lucid Common Lisp 4.0), Version 3.2 on MacIntosh (MacIntosh Common Lisp 2.0) and Version 2.2 on VAX/VMS (VAX Common Lisp and UTC-LISP). For version 2.2 on UTC-LISP, there is an object server MGF [Barthès 85]. And for version 3.2 on MacIntosh, we use a simulated database for keeping all objects in a sequential file because the MacIntosh environment does not have a server for sharing objects among several users. For Version 3.1, we are currently using MATISSE™[1] for storing objects. We are also developing another new object server for MOSS.

4 A Small Example in Mechanical CAD

Here is a small example of an application of the object-oriented programming environment MOSS in the field of mechanical CAD.

4.1 Representation of mechanical objects

Because of the important features of MOSS such as semantic representation, versioning, persistency, concurrence, MOSS is quite suitable for the representation

[1] MATISSE is a trademark of ADB

of mechanical objects [Shen et al 94], the representation of design knowledge [Finger & Rinderle 89], and even the representation of messages and agents in the distributed integrated design environment [Shen & Barthès 94].

We now explain how we use MOSS in our design environment for representing mechanical parts and for propagating constraints.

We consider a simple mechanical assembly consisting of two plates fixed by means of a bolt, a nut and a washer (see Fig. 3). This assembly can evidently be represented as in Fig.4 using hierarchical relations, and can also be represented completely as in Fig.5 using semantic relations.

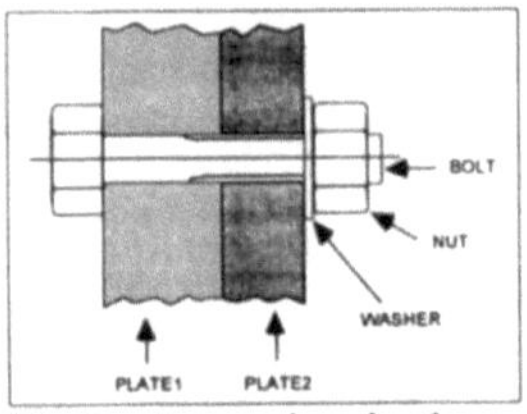

Fig.3 A mechanical assembly

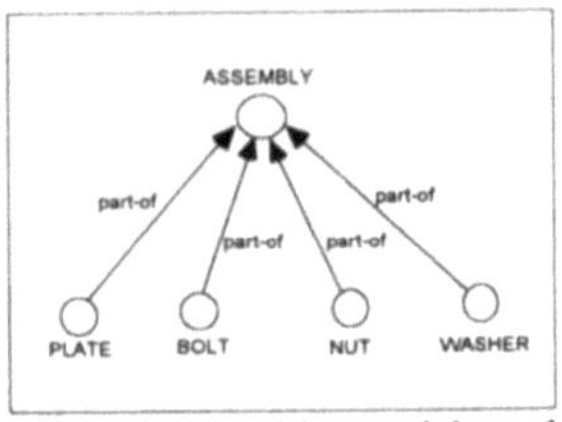

Fig.4 The hierarchic relations among the classes

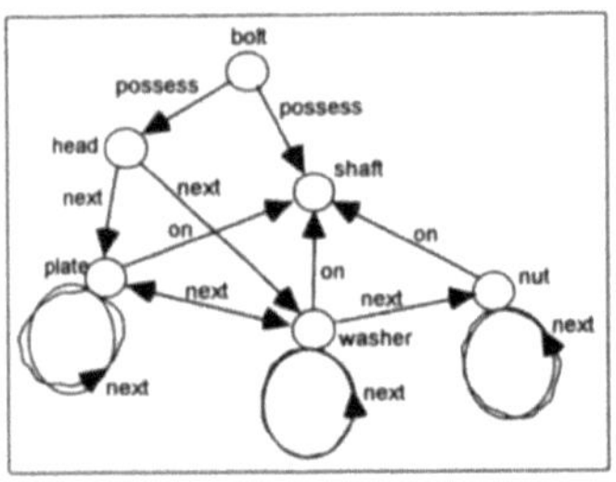

Figure 5. The semantic relations among the classes

The definition interface of MOSS allows us to create classes simply. For example, the class WASHER can be defined as follows:

```
(defclass WASHER $WASH test
        (:sp PART-OF ASSEMBLY)
        (:tp THICKNESS)
        (:tp INTERNAL-DIAMETER)
        (:tp EXTERNAL-DIAMETER)
        (:sp WASHER-TYPE WASHER-TYPE)
        (:sp NEXT NUT WASHER))
```

A washer is defined with three terminal properties: thickness, internal diameter (standard) and external diameter (standard); it is related to other classes through the structural properties: washer type (correspondent to a standard table), composition relation PART-OF and the possible relation NEXT with the matching parts.

4.2 Constraint propagation in mechanical CAD systems

We have defined several methods for modifying properties of objects and for propagating constraints. We have compared the two techniques for constraint propagation, namely, along the hierarchical link IS-PART-OF to a higher level of assembly, and along a local semantic link such as NEXT to the matching parts. We selected the latter for our system because it allows the components of the system to be presented clearly by using semantic links and because constraint verification can be made while the objects are being modified.

If we have to modify this mechanical assembly, e.g. we decide to modify the bolt diameter because it is too thick or too thin, then, it can be done simply by sending a message:

```
(send <bolt> '=modify-diameter <new-value>)
```

In the system, the modification propagates automatically to the "next" part by a new message contained in the method *=modify-diameter* for the class BOLT:

```
(send (g==> 'HAS-NEXT '=modify-value <Φ   >)).
```

Propagation continues as new messages are sent to other objects.

We have used the object oriented programming environment MOSS for representing more complex systems such as a gearbox [Shen et al 94].

CONCLUSION

Most industrial applications need to handle a large amount of information which is linked by complex relationships. The traditional programming environment and database technology cannot meet such complexity. An object-oriented representation model, PDM, was proposed in this situation. An object-oriented programming environment MOSS was developed as a research prototype for implementing all the features of the last version of the PDM model. This object-oriented programming environment MOSS has several important features such as persistency, versioning, concurrence control, in addition to the general features of the object-oriented system.

A small example has been presented to demonstrate the application of MOSS in the field of mechanical CAD systems, which proved that MOSS is adequate for developing CAD applications, and that object-oriented technology such as object-oriented programming and object-oriented database technology provides an effective solution for complex engineering problems.

PDM is an object representation model in development and its implementation MOSS described here refers to a research prototype. The commercial implementations of PDM such as G-BASE™, MATISSE™ use simpler mechanisms. It was found in that case that a few million objects could be handled easily.

However, there are still some problems unsolved, such as persistency, query mechanisms, natural language interface, which are the perspectives of our future work.

KEYWORDS

Artificial Intelligence, object-Oriented, representation model, programming Language, mechanical CAD.

REFERENCES

[Abrial 74] Abrial J.R., Data semantics, in Data Base Management, Klimbie J.W., Koffeman K.L (eds), North-Holland, pp 1-59 (1974).

[Barthès et al 79] Barthès J. P., Vayssade M., Miaczynska-Znamierowska M., Property Driven Data Base, 6th IJCAI, Tokyo (1979).

[Barthès 85] Barthès J.P., MGF Version 2.0M: Moniteur de Gestion de Fichiers, UTC/GI/DI/85-12, Dec.(1985).

[Barthès 89] Barthès J. P., MOSS: a multi-user object environment, 2nd Symposium on AI, Monterrey, Mexico, Oct. (1989).

[Barthès et al 90] Barthès J.P., Anota P., El Dahshan K., An Experience in Adding Persistence to Intelligent CAD Environments, in Intelligent CAD Systems, ten Hagen P.J.W., Veerkamp P.J. (eds), Springer Verlag (1990).

[Barthès 93] Barthès J.P., La problèmatique de reconciliation en ingénierie simultanée, Actes de 01 DESIGN '93, Tunis, Nov.(1993).

[Bobrow et al 77] Bobraw D.G., Winograd T., An overview of KRL, a Knowledge Representation Language, Cognitive Science, 1(1), pp 264-285 (1977).

[Clayton 85] Clayton B.D., ART: Reference Manual, Inference Corporation, Los Angeles, CA (1985).

[Fox et al 86] Fox M., Wright J.M., Adam D., Experiences with SRL: An analysis of a frame-based knowledge representation, in Expert Database Systems, Kerschberg L. (ed), TheBenjamin/Cummings Publishing Company Inc., pp 161-172 (1986).

[Greiner et al 80] Greiner R., Lenat D.B., A Representation Language Language, Proc. AAAI-80Stanford, pp165-169 (1980).

[Kim et al 87] Kim W., Banerjee J., Chou H., Garza J., Woelk D., Composite object support in an object-oriented database system, in OPS 87, pp 118-125 (1987).

[Leuna et al 90] Lenat D.B., Guha A., Building Large Knowledge-Based Systems, in Reading, Addison-Wesley (1990).

[Lécluse et al 87] Lécluse C., Richard Ph., Velez F., O2, An Object-Oriented Data Model, Rapport ALTAIR, Rocquencourt, France (1987).

[Minsky 75] Minsky M., A framework for representing knowledge, the psychology of computer vision, (ed)Winston PE., MacGraw Hill (1975).

[Schank 91] Shank R.C., Where's the AI?, in AI Magazine,12(4), pp 38-49 (1991).

[Shen et al 94] Shen W., Barthès J.P., EL Dahshan K., Propagation de Contraintes dans les Systèmes de CAO en Mécanique, Revue internationale de CFAO et d'infographie, Volume 9(1-2), pp 25-40, (1994).

[Shen et al 94] Shen W., Barthès J.P., Toward a multi-agent architecture for Distributed Integrated Design Environment (DIDE), Tech. Report 94-1, CNRS URA 817 Heudiasyc, UTC, Jan.(1994).

[Finger et al 89] Finger S., Rinderle J.R., Representation of Mechanical Design, The Third Workshop on Intelligent CAD, IFIP Working Group 5.2, Osaka, Japon, Sep.(1989).

[Yu et al 86] Yu X., Ohbo N., Masuda T., Fujiwara Y., Database Support for Solid Modeling, in The Visual Computer, 12(6), pp 358-366 (1986).

A STEP TOWARDS THE CLIENT/SERVER MODEL: THE OODE APPROACH

Michel RIVEILL[1] *and Gérard VANDOME*[2]

1 Université de Savoie - LGIS Campus scientifique F-73370 Le Bourget du Lac, France E-mail : Michel.Riveill@univ-savoie.fr
2 Bull/Open Systems and Software 2, rue de Vignate F-38610 Gières, France E-mail: G.Vandome@frgu.bull.fr

ABSTRACT

This paper describes the main functionalities of the OODE (Object Oriented Distributed Environment) platform which facilitates the development of distributed and cooperative applications using persistent objects. The compatibility with the OMG and CORBA is also discussed.

RESUME

Ce papier décrit les principales fonctionalités de la plate-forme OODE dont le but est de faciliter le développement d'applications distribuées et coopératives utilisant des objets persistants. La compatibilité avec l'OMG et CORBA est également discutée.

1 INTRODUCTION

The objective of the OODE (Object Oriented Distributed Environment) platform is to facilitate the development and to support the execution of cooperative applications using persistent and distributed objects. OODE is derived from a prototype implementation achieved within the framework of the CoManDOS (Construction and Management of Distributed Open Systems) Esprit project [1]. It is currently used to develop cooperative applications such as the CIDRIA application developed by the PREVISIA Consortium. CIDRIA is a distributed workflow application. PREVISIA (Distributed Platform for Evaluating and Integrating Advanced Information Systems) is a consortium formed by BNP, EDF, France-Télecom/CNET and INRIA in order to experiment and use new technologies.

The OODE project has adopted an innovative approach using integration of programming language and operating system technologies. A unified view is provided by a model and a system architecture based on an Object Oriented technology coupled with distributed persistent storage. The object paradigm gives an integrated view of application design, programming, management and control.

OODE main functionalities are described in section 2. The corresponding development environment is briefly presented in section 3, especially how OODE functionalities are provided to the programmer. Section 4 compares OODE and Object Management Group (OMG) approaches, how OODE offers advanced functionalities while remaining compliant with the OMG CORBA (Common Object Request Broker Architecture).

2 OODE MAIN FUNCTIONALITIES

OODE's conceptual model encompasses object distribution and persistence. It presents the OODE platform functionality to application designers, programmers and administrators. The main functionalities of the OODE Object Model are :

- A distributed execution model where the distribution is fully transparent : a remote object invocation is expressed by the programmer as a local one;
- Persistent programming : object loading (from secondary storage to virtual memory) and object saving (from virtual memory to secondary storage) is done automatically by the OODE platform when necessary (the programmer need not take care of files and data storing and retrieving);
- The ability to share objects in parallel among several applications, in a "What You See is What I see" procedure;
- Object Oriented protection and security mechanisms.

2.1 Distributed execution model

The OODE object is the unit managed by the system. It is characterized by its data and a set of operations. OODE objects are passive and are activated within specific execution structures. The unit of execution, *job*, gives the programmer a virtual machine where parallelism is visible and distribution is hidden. A job, usually can consist of several sequential distributed threads of control called *activities*. The job objects are dynamically loaded and linked within a distributed execution space called context. The OODE execution model generalizes a set of nodes of the traditional model of process and threads now provided by most operating systems. OODE gives the programmer this uniform model (distributed processes and threads) which replaces the client/server model where clients and servers require specific programming. The OODE programmer does not have to deal with distribution aspects and can concentrate all his effort on the application. The distribution is defined afterwards by configuring the application (and not by programming). The distribution can be controlled by policies defining where objects should be stored, executed, etc. that can be defined and modified by administrative tools.

2.2 Persistent programming

An object is said to be persistent if its lifetime is not bound to a given program execution. In conventional programming languages the lifetimes of entities are bound to the duration of the program run. Files are usually the sole exception, with the consequence that if a particular entity is to survive a program execution, it must be inserted into a file (possibly requiring a change in its representation) and explicitly retrieved (and possibly rebuilt) at a later time. This requires (when an object has to be stored) the conversion, by the programmer, of all memory pointers to other persistent objects into something meaningful within secondary storage (and vice-versa when the object has to be retrieved).

In OODE, persistence is defined via the reachability property of objects. This means that every object reachable by recursively navigating through the component objects, from some specified set of root objects, is guaranteed by the system to persist. Thus no dangling reference to objects may exist. When not is use persistent objects are stored in secondary storage. When an attempt to access such an object is made, the object is automatically fetched from storage by the system and made available in virtual memory at an appropriate node. Hence it is unnecessary for the programmer to write code either to explicitly fetch the object from storage or to convert between the possibly different secondary and virtual memory representations of the object. Likewise, when an object is no longer required, the system automatically stores the object back to secondary storage: there is no need for the programmer to explicitly save the object.

When an object is retrieved from storage, other related objects may be retrieved at the same time and in the same I/O operation so that they are immediately available for use when needed. Such a group of related objects is called *cluster of objects*. This improves performances by reducing the number of mapping and I/O operations required when related objects are used together.

An object is identified by a distributed pointer allowing unique identification and location. This pointer is valid both in virtual memory and in secondary storage. The storage policy (where the object is stored) is defined independently of the execution policy. The state of an object can contain distributed pointers to other remote objects, thus allowing to build compound objects whose components may be on several nodes.

2.3 Object sharing

Using the client/server model, programming (and compilation process) of the client and the server are different. The invocation of an object is different if it is called from an object located within the same address space (for example invocation of a C++ object) or within another address space (object managed by a server). The invoked object is located within a different address space than the invoking object, and it is thus not possible to share any data or objects between the client and the server objects. Server object programming is still more different as it requires a set of specific procedures before being able to be called as a server object.

With OODE all objects are programmed and compiled in the same way, independently of the fact that they can or cannot be invoked by another object. Any object can invoke or be invoked by any other object, exactly as within C++, but this is no longer restricted to a single address space and temporary objects. The same object can be simultaneously used by several applications without requiring an intermediate server process. This is made possible by the ability to share a given object between several address spaces. This is similar to the classical shared memory provided by UNIX, but this is not restricted to a single node, and is automatically done at run-time (while the UNIX shared memory has to be explicitly managed by the programmer).

Object sharing is particularly adequate for cooperative applications which have to share common data between several users. It is easier to use than explicit communication through a server and it provides real simultaneous data sharing service also called WYSIWIS ("What You See Is What I See").

2.4 Security and protection

Maintaining the security of a distributed object system entails maintaining object security and protection.

- *security* means objects cannot be corrupted (maliciously or because of a programming error);
- *protection* means that the objects are accessible only to authorized users.

2.4.1 Security

In the client/server approach, to provide minimum execution security the server creates one dedicated address space for each client so that server failure does not induce the failure of all its clients. If the server only creates a thread for each client, it does not provide any security for client execution as all threads use a common address space which can be corrupted by any of them.

With OODE there is at least one distributed address space per job (session of an application for a given user): this provides a complete isolation between running applications. Only objects which are simultaneously used by several running applications (jobs) are shared between corresponding address spaces. An additional security level is provided by the creation for each job, of one address space for each owner of the objects manipulated by the corresponding application: only objects belonging to the same user are loaded within a given address space. This ensures, in a way transparent to the programmer, a security level difficult to obtain with a client/server approach: an object belonging to a given user cannot be corrupted by an object belonging to another user.

2.4.2 Protection

The protection of objects manipulated by the server is only provided by access rights on files in which these objects are stored. The expression of access rights to objects is often limited to read only, or read and write for a given user or group of users (as for UNIX files). Save for specific mechanisms managed by the server (as with databases for example) the protection must be the same for all objects managed by a given server. Read and write access rights are often inadequate for object programming where data are encapsulated and are only accessible through methods: for example it can be necessary to restrict to a given subset the methods that a user is allowed to invoke on a given object.

With OODE it is possible to define views. A view on a class is a subset of all the possible methods of this class. User rights are defined by views on objects, and no longer use read and write access rights. If a user has the right to a given view on an object, it means that he can only invoke methods associated to this view. For example the *secretary* view on the *employee* object representing Jim can be allocated to Jane (Jim's secretary) and the *manager* view to John (Jim's Manager). Both views can give (through methods) the ability to modify different fields encapsulated within the employee object: for example Jane can modify Jim's address and John, Jim's salary. This provides a very flexible mechanism for access rights definition and control.

3 Development Environment

The OODE development environment is composed of a language called OC++ (for OODE C++) and of a set of associated tools. There are two main ways to extend C++, which are, either to develop specific classes which provide the required extensions by inheritance, or to extend the syntax of the language. For flexibility and coherence reasons, which cannot be detailed here, we have chosen the second approach by adding a single keyword "distributed" to the C++ syntax. OC++ is an extension to the standard C++ which provides adequate support for the programming of distributed applications using persistent data. This language is detailed in [2]. Only the part of the application using the distribution and persistency is impacted by the new aspects of the language. Some C++ features, as friends and use of non distributed pointers, are forbidden within distributed classes.The OC++ language is processed by a preprocessor which performs all needed semantic and syntactic checks and generates standard C++ code, which is then processed without any change by the standard C++ compiler of the target machine. Distributed objects (both data and corresponding code) are dynamically loaded and linked at runtime.

4 OODE and CORBA

The OMG has defined an Object Model and an Architecture [4]. The OMG is an organization regrouping more than 400 system vendors, ISVs and users, whose aim is to define standards for Distributed Object Management. The ORB ("Object Request Broker") is the core component of the OMG Architecture. It allows objects to exchange requests among themselves. The architecture of the ORB component, called CORBA is defined within. One of the main contributions of this standard is the definition of the IDL ("Interface Definition Language") which allows one to define an object interface of an object independently of the language which is used to implement it.

The OODE Object Model and Architecture are compliant with the Object Model and the Architecture defined by the OMG. The interface to OODE objects is defined using the OMG IDL; this allows the invocation of any OODE object through CORBA. Any OODE object can also call any other object through CORBA. Invocation between OODE objects however use specific optimized mechanisms which, in addition, provide the additional functionalities detailed in section 2 (distributed execution model, object persistency, object sharing, security and protection), with the same level of performance as for standard C++ objects. The ability to invoke external objects through CORBA is used, for example, to access objects stored within object and relational databases. An additional tool, described in [6], allows one to give an object view of data stored within relational databases, by generating from the database schema, C++ classes encapsulating relational data.

Conclusion

This experience has shown the main benefits expected from the use of the OODE platform for the development of new distributed applications and for the distribution of existing centralized applications which are redesigned for a distributed environment.

Keywords

Object-oriented, concurrent, distribution, persistence.

References

[1] Cahill V., Balter R., Harris N.R., Rousset de Pina X, The Comandos Distributed Application Platform, Springer-Verlag (1993).

[2] Santana M., OC++: A proposal for extending C++ within ASSET, Dec.(1993).

[3] The Common Object Request Broker Architecture and Specification, Object Management Group Inc,, 492 Old Connecticut Path, Framingham, MA 01701, Dec.(1991).

[4] Soley Richard Mark, Second edition of the Object Management Architecture Guide, containing updated OMG Object Model., Object Management Group Inc,, 492 Old Connecticut Path, Framingham, MA 01701, Nov. (1992).

[5] Thompson Craig , Object Services Architecture - Revision 6.0, Object Management Group Inc, 492 Old Connecticut Path, Framingham, MA 01701, Aug. (1992).

[6] Exertier F., Generating OO Interfaces for Relational Databases, Apr.(1994).

FROM BIBLIOGRAPHIC DATA BASES TO KNOWLEDGE BASES

Nicolaas MARS[1]

Knowledge-based Systems Group, Department of Computer Science, University of Twente, P.O. Box 217, 7500 AE Enschede, The Netherlands

ABSTRACT

We describe the Plinius project, carried out at the University of Twente. The aim of this project is to automate partially the process of extracting knowledge from machine-readable natural-language text, in particular title and abstract of scientific journal articles. A case study in the domain of mechanical properties of ceramic materials is currently being carried out. The engineering decisions made to make the approach feasible are explained.

RESUME

Nous décrivons le projet Plinius développé à l'Université de Twente. L'objectif de ce projet est d'automatiser le processus d'extraction de connaissances à partir d'un texte informatisé en langage naturel, en particulier à partir des titres et résumés des articles de journaux scientifiques. Une étude de cas est en cours sur les propriétés mécaniques des matériaux céramiques. Les décisions de développement d'ingénierie réalisable sont expliquées .

1 INTRODUCTION

Large amounts of scientific knowledge and data are currently available in bibliographic data bases. However, even though this material is in machine-readable form, it is not suitable for direct processing in computer systems. Intervention of humans, with considerable expertise in the domain involved, is needed to extract knowledge from the texts and encode it in a machine representation.

The costs involved in knowledge acquisition are generally considered to be an important impediment to the wide-spread introduction of knowledge-based systems.

---

1 This article describes joint work of H. de Jong, F. van Raalte, P.-H. Speel, W.G. ter Stal, P.E. van der Vet, and the author, all at the University of Twente, Enschede, The Netherlands.

One proposed solution is the acquisition of knowledge by computer analysis of natural-language text. Over the past two decades several projects for knowledge extraction from texts have been reported [1, 2, 3, 4]. We compiled an overview [5] of twenty such projects. However, few projects seem to have proceeded beyond the stage of preliminary investigations and/or small prototypes.

The Plinius project at the University of Twente aims to partially automate the process of extracting knowledge from machine-readable texts. In particular, the project is aimed at showing the feasibility of extracting an economically interesting subset of the knowledge present in the abstracts of scientific journal articles. Experiments are being done using the Engineered Materials Abstracts (EMA)[2]. The aim of the Plinius project is to realize a semi-automatic knowledge acquisition system which produces an economically valuable knowledge base. In order to attain this goal, we have decided to pursue an engineering approach. An engineering approach means adopting a pragmatic stance towards the (sometimes fundamental) problems encountered during the project. We have attempted to solve or avoid these problems with the economic viability of the final result in mind.

2 Overview of the Plinius Process

The analysis process consists of several stages. In a preprocessing stage, recognition of sentence boundaries, formulae, etc. is done. Then, in the language-dependent process, a syntactic, semantic, and discourse analysis is carried out on the preprocessed abstracts, and this leads to the representation in a formal language of the knowledge contained in the documents. These formal representations are stored in the interim knowledge base. Knowledge bases used in this stage are domain-dependent but language-independent. Finally, in the language-independent process the interim knowledge base forms the basis for an integration procedure, which amounts to the integration of the knowledge elements extracted from different abstracts. The language-independent process makes use of domain-specific knowledge sources, including an ontology. The ontology prescribes the concepts and relations, defined as important in the domain.

We do not intend to come up with a fully automatic system in the Plinius project. All three processes can communicate with a human operator who deals with natural language ambiguities and other problems the system cannot solve automatically. The interim and integrated knowledge bases are the final products of the Plinius project; they can be put to use in a wide range of applications. For example, the interim

2 Materials Information is a registered trademark of The Institute of Materials (London) and ASM International (Metals Park OH). EMA collects all abstracts about materials other than metals.

knowledge base might be used in a system for semi-automatic indexing. The integrated knowledge base is meant to be a general-purpose, domain-specific knowledge base and might be used for directly answering user queries. In the ideal case, all user queries that can be answered in terms of our ontology by reading the abstracts are covered by the integrated knowledge base as well. In addition, the integrated knowledge base could be combined with other domain-specific knowledge bases to arrive at systems dealing with a larger domain.

To date we have completed the following components within the project. The preprocess has been realized; this means that redundant information in the abstract is eliminated, formulae are identified and reformatted, and abbreviations ending with periods are distinguished from end of sentence markers. The grammar needed for the syntactic and semantic analysis of sentences is written in patr-II, a unification-based grammar formalism. Currently, the grammar consists of about 35 rules covering basic linguistic constructions occurring in our corpus. We use a fast head-corner (chart) parser to analyze input sentences.

The output of the parser is twofold: a) a parse tree graphically displaying the syntactical structure of the sentence and b) a feature structure containing both syntactic and semantic information. Semantic information is expressed in terms of a canonical predicate-argument structure which forms the input for further processing. The ontology currently consists of a number of atomic concepts and construction rules, covering ceramic materials, their properties, and the processes to make them. The ontology is described elsewhere.[6] The ontology has been formalized and implemented in several KR languages, such as classic, loom and back. The idea is to evaluate these implementations against a number of criteria and to make an adequate choice for an appropriate language. Current research consists of developing a method to associate the semantic information obtained from the parser with the ontology KB, the development of the discourse process, the design of the background KB and a detailed investigation into the problem of interpreting compound nouns.

3 THE ENGINEERING APPROACH IN PLINIUS

The most important design decisions we have taken in the Plinius project are the use of abstracts, the restriction to a sublanguage, the interactive resolution of complex problems, the application of an ontology, the reuse of linguistic and knowledge engineering resources, and the permitted occurrence of imperfections. Although individual design decisions may have been taken in other projects as well, we believe that the combination of these decisions informed by a clear economical goal is unique and distinguishes Plinius from the other projects with similar goals. In our view, the adoption of an engineering view may lead to success where other semi-automatic knowledge acquisition projects failed.

Our focus on engineering aspects does not imply that the results of the Plinius project are without scientific value. The insights obtained from completing a realistic application may contribute to major research issues in the fields of NLP, knowledge engineering, and information retrieval. In fact, as spin-off the Plinius project has yielded a general ontology for part of the chemical domain [6] and a method to resolve inconsistencies in scientific knowledge bases [7].

3.1 Use of abstracts

Each year, Materials Information produces thousands of abstracts of scientific publications in the field of materials science, called Engineered Materials Abstracts (EMA). As input to the Plinius knowledge extraction process, the EMA database is used containing the document descriptions in machine-readable form. We have selected a corpus of 400 abstracts on mechanical properties of ceramic materials. By their contents, these abstracts implicitly define the domain of Plinius. The corpus has been split into two parts: the 300 abstracts dating from the first three quarters of 1990 aid in the development of the system and the remaining 100 abstracts from the last quarter of 1990 will eventually be used to test the system. An example of a part of an EMA document description is given below.

ID : *9005C1-C-1123*

TI : *Superplasticity of Hot Isostatically Pressed Hydroxyapatite.*

AU : *F. Wakai, Y. Kodama, S. Sakaguchi, T. Nonami Government*

Industrial Research Institute (Japan), TDK

AB : *Dense and translucent hydroxyapatite polycrystals* $Ca_{10}(PO_4)_6(OH)_2$ *with a grain size of 0.64 u m) were obtained by hot isostatic pressing at 203 MPa and1000 °C for 2 h in Ar. The material exhibited superplastic elongation (> 150%) in a tension test at temperatures from 1000-1100 °C and at strain rates from* $7.2x10^{-5}$ to $3.6x10^{-4}$ s^{-1}. *Extensive strain hardening was observed. The stress exponent of the yield stress was larger than three.*

SO : *GJ. Am. Ceram. Soc.IFeb. 1990, J73, (2)$L457-460 MEnglish QISSN 0002-7820*

IN : *Hydroxyapatite, Mechanical properties Superplasticity Hot isostatic pressing Deformation*

Part of a document description of EMA, consisting of identifier (**ID**), title (**TI**), author(s) (**AU**), abstract (**AB**), source (**SO**) and index terms (**IN**).

The use of abstracts instead of full-text articles is motivated by a number of reasons. Apart from the obvious technical advantage of the direct availability of machine-readable input, the abstracts in the EMA database are presented in a standardized format which simplifies the preprocessing task. In addition, abstracts are concise and well-structured, contain no illustrations, jokes or metaphors, and are written in a uniform style. As a consequence, certain interpretation problems in the language-dependent process are relaxed. The question can be raised whether the abstracts contain enough information for the purpose of constructing a knowledge base which is capable of answering a satisfactory proportion of user queries. This question becomes even more pressing when it is realized that semi-automatic knowledge acquisition will be able to extract only a part of the knowledge conveyed by the abstract.

Our belief that the abstracts nevertheless present an economically valuable subset of the knowledge available in a domain is based on three considerations. In the first place, due to the large number of publications in the field of materials science, many users try to keep themselves informed by reading relevant titles and abstracts only. From this it might be conjectured that the abstracts in EMA contain a sizable proportion of the relevant knowledge. As a second consideration, it must be recalled that in Plinius the knowledge is drawn from a large collection of abstracts, so that knowledge which is not mentioned in one abstract may well be found in other abstracts. The integration of knowledge derived from different abstracts may fill up the lacunae that originate from the loss of information in writing the abstract or in semi-automatically extracting the knowledge. A final reason for assuming that the knowledge base acquired from the abstracts is sufficiently informative, derives from the possibility to draw inferences. Knowledge which is implied by the abstracts, though not explicitly mentioned, can be inferred from the knowledge base by the reasoning mechanism included in a knowledge representation system.

3.2 Restriction to sublanguage

One of the main engineering advantages of the language-dependent process in Plinius is that it is not expected to deal with the English language in all its richness and complexity. Instead we are developing a process tailored to a more comprehensive subset of English used in the abstracts of the Plinius domain. This subset of English is restricted compared to the English language as a whole. The motivation for employing this so-called {sublanguage} is that automatic analysis of unrestricted text in arbitrary domains is not (yet) possible. However, due to the restrictions at the various linguistic levels, automatic linguistic analysis seems feasible.

There are also a few disadvantages of the Plinius sublanguage. Due to the very nature of abstracts, they contain information in a condensed format. Linguistically, this means that constructions considered generally as problematic, such as noun phrases containing several prepositional phrases, noun-noun compounds and coordinated phrases occur frequently in our corpus.

3.3 Interactive resolution of complex problems

In the previous section we argued that employing a sublanguage reduces the complexity of the task of developing the language-dependent process for the Plinius purpose. However, despite this reduction we expect some problems to remain unsolved. Examples of these problems are :

> **[Lexical incompleteness]** The language-dependent process will have to deal with unknown words in one way or another.

[**Structural ambiguities**] Although we are fairly optimistic about the ability to resolve structural ambiguities by means of selection restrictions, there will always be a limited number of sentences which yield multiple analyses.

[**Anaphoric resolution**] There will be cases in which the discourse process will not be able to establish the correct anaphoric/antecedent relations.

One way of dealing with the alleged deficiencies of the system is to develop more sophisticated NLP modules capable of resolving these problems. However, we expect there will be a negative trade-off between the amount of research yielding these elaborated modules and the improved performance of the system. Therefore, we opted for a typical engineering solution by allowing the system to interact with the user. Thus, the aforementioned problems are being tackled in a semi-automatic fashion while awaiting results from research within computational linguistics to deal with them automatically. Of course, attempts will be made to minimize the amount of interaction with a human operator.

3.4 Application of an ontology

The Plinius project is concerned with transforming sentences in natural language into assertions expressed in a knowledge representation language. In order to support this conversion process we have decided to develop an ontology for the Plinius domain. An ontology, a term introduced in this sense by Hayes [8], consists of a vocabulary of unambiguous concepts and their interrelations. An ontology presents a framework which structures the domain of interest at the knowledge level. As a consequence, the development of an ontology proceeds independently from representational and implementational considerations. The ontology used in the Plinius project consists of concepts that can be employed to express relevant knowledge on ceramic materials, such as their chemical composition, their (mechanical) properties, and the processes to produce them. The Plinius ontology can be considered a conceptual construction kit consisting of pre-defined atomic concepts and construction rules for defining composite concepts in terms of atomic concepts. The atomic concepts and construction rules together implicitly define a large number of concepts. A detailed overview of the Plinius ontology is available.[6]

The ontology facilitates the conversion from natural language into a knowledge representation formalism in a number of ways. In the first place, the words and phrases occurring in the abstracts are mapped onto concepts in the ontology. This mapping is achieved through the lexicon in which ontological concepts are attached to the words and phrases making up the lexical entries. Thus, the ontology can be considered a specification of the vague notion of the semantics of natural language. A practical consequence of defining the meaning of natural language expressions in terms of concepts in the ontology, is that it aids in the resolution of lexical ambiguities. The natural language terms `alumina',`aluminium oxide', `aluminum oxide', and `Al_2O_3' all refer to the same chemical concept, namely that for the compound identified. In general, a single concept in an ontology corresponds to different words and phrases, both within a single natural language and in different natural languages.

Further ways in which the ontology assists in the process of knowledge acquisition concern the facilitation of coordination between the different knowledge bases used as resource by the system and the predefinition of a clear structure for the output knowledge base. In addition, the (declarative) semantics of the assertions in the knowledge base can be specified by means of an interpretation function linking the non-logical symbols in the knowledge representation language to concepts in the ontology. Apart from the advantages in actually building the Plinius system, the use of an ontology carries an advantage regarding the further utilization of the output knowledge base, e.g., in systems for materials selection or in systems that combine the output knowledge base with existing databases and knowledge bases. These favorable conditions for the reuse and sharing of knowledge increase the economic attractiveness of the Plinius knowledge base.

3.5 Reusing linguistic and knowledge engineering resources

Another engineering characteristic of the Plinius project is the choice for reusing, as much as possible, existing linguistic and knowledge engineering resources. Here, the term resources stands for a number of notions.

First of all, it designates implemented formalisms such as classic for knowledge representation and patr-II, a unification-based grammar formalism for the specification of NL grammars.

Secondly, it refers to broad-coverage NL grammars such as Alvey, and machine-readable lexicons such as the OALD. Specifically, we use the Alvey grammar as an inspiration source for deciding how to specify grammar rules for certain grammatical constructions.

Although we aim at reusing existing resources as much as possible, in practice this idea sometimes turns out to be rather cumbersome. Due to certain idiosyncrasies and lack of standards, tools and formalisms can not be reused directly.

3.6 Permitted occurrence of imperfections

The design decisions discussed in the previous sections may give rise to the occurrence of imperfections in the output knowledge base. Although we strive at reducing imperfections as much as possible, within certain bounds incomplete and incorrect knowledge does not render the output knowledge base useless. Incompleteness, the absence of available domain knowledge in the output knowledge base, arises from the use of abstracts instead of full-text articles and the fact that only knowledge expressible in terms of concepts from the ontology appears in the output of the knowledge acquisition process. In addition, the use of a sublanguage grammar may lead to a loss of information. Incompleteness may be partly overcome by the redundancy expected to occur in a large corpus of abstracts. The occurrence of erroneous knowledge in the interim knowledge base, incorrectness, arises from errors in the abstracts and from inadequacies in the process of mapping natural language sentences to ontological concepts. Part of the errors will manifest themselves as inconsistencies and will subsequently be handled by the language-independent process.

CONCLUSION

The Plinius project is characterized, perhaps more than other semi-automatic knowledge acquisition projects, by an engineering approach. This engineering approach is reflected in the design decisions taken, such as the use of abstracts, the employment of a sublanguage, the interactive resolution of difficult linguistic problems, the application of an ontology, the reuse of linguistic and knowledge engineering resources, and the permitted occurrence of imperfections.

KEYWORDS

Knowledge base, knowledge extraction, knowledge acquisition, natural-language analysis.

REFERENCES

[1] Herzog O., Rollinger C.-R., Text understanding in LILOG: integrating computational linguistics and artificial intelligence, in Lecture notes in artificial intelligence, vol. 546, Springer-Verlag, Berlin (1991).

[2] Antonacci F. & al , A system for text analysis and lexical knowledge acquisition, in Data and Knowledge Engineering, 4, 1-20 (1989).

[3] Rau L.F., Jacobs P.S., NL and IR: natural language for information retrieval, in International journal of intelligent systems,4, 319-343 (1989).

[4] Hahn U., Making understanders out of parsers: semantically driven parsing as a key concept for realistic text understanding applications, in International Journal of Intelligent Systems, 4, 345-393 (1989).

[5] Mars N.J.I., Knowledge acquisition from natural-language texts, A review of the literature.University of Twente, UT-KBS-91-23, Enschede, The Netherlands (1991) (updated version to appear soon).

[6] van der Vet P.E., Mars N.J.I., Structured system of concepts for storing, retrieving, and manipulating chemical information, in Journal of Chemical Information and Computer Sciences, 33, 564-568 (1993).

[7] de Jong H., Mars N.J.I., van der Vet P.E., Consistency of scientific knowledge bases, in Working notes of the AAAI93-workshop on Validation and Verification of knowledge-based systems,A.D. Preece et al. (eds.), AAAI, 64-65 (1993).

[8] Hayes P.J., Naive physics I: ontology for liquids, in Formal theories of the commonsense world, Hobbs J.R. , Moore R.C.(eds.), Ablex, Norwood, NJ, 71-107 (1985).

Chapter 2

IMPRECISION AND UNCERTAINTY IN INFORMATION SYSTEMS

IMPRECISION AND UNCERTAINTY

Philippe SMETS

IRIDIA, Université Libre de Bruxelles, 50 avenue Franklin D. Roosevelt, CP 194/6, 1050 Bruxelles, Belgium

ABSTRACT

This report shows the difference between imprecision or uncertainty, and the models that have been proposed to represent them: fuzzy sets theory, possibility theory, probability theory and the transferable belief model.

RESUME

Ce texte présente la différence entre l'imprécis et l'incertain, et les modèles proposés pour les représenter: la théorie des ensembles flous, la théorie des possibilités, la théorie des probabilités et le modèle des croyances transférables.

1 IMPRECISION AND UNCERTAINTY

Imperfection, be it imprecision or uncertainty, pervades the real world and must be incorporated into every information system. Yet, this is hardly achieved today. A major reason might be found in the difficulty in understanding the various aspects of imprecision and uncertainty.

Until recently, almost all aspects of imperfect data were modeled by the probability theory, but in the last 20 years, many new models have been developed to represent imperfect data. The large number of models reflects the recent acknowledgment that there exist many aspects of imperfection and that the probability theory, as good as it is, is not the unique normative model that can cope with all of them.

This paper will present some aspects of imprecision or uncertainty. We present some of the approaches that have been proposed to model their quantified forms. A detailed analysis is presented in Smithson (1989) and Smets (1991).

2 Variety of Imperfect Information

Information is perfect when it is precise and certain. Imperfection can be due to imprecision, inconsistency and uncertainty, the major aspects of imperfect data. Imprecision and inconsistency are properties related to the content of the statement: either more than one world or no world is compatible with the available information, respectively.

Imprecision and inconsistency are essentially properties of the information itself whereas uncertainty concerns the state of knowledge of an agent about that relation between the world and the statement about the world.

The statement itself is either true or false, but the agent's knowledge about the world does not allow him to decide definitively if the statement is true or false. Certainty is full knowledge of the true value of the data. Uncertainty is partial knowledge of the true value of the data. Uncertainty is essentially, if not always, an epistemic property induced by a lack of information. A major cause of uncertainty is imprecision in the data. Whether uncertainty is an objective or a subjective property is a still debated philosophical question left aside here.

To illustrate the difference between imprecision and uncertainty, consider the following two situations :

1. John has at least two children and I am sure about it.
2. John has three children but I am not sure about it.

In case **1**, the number of children is imprecise but certain.

In case **2**, the number of children is precise but uncertain. Both aspects can coexist but are distinct.

Often the more imprecise you are, the most certain you are, and the more precise, the less certain. There seems to be some Information Maximality Principle that requires that the 'product' of precision and certainty cannot be beyond a certain critical level. Any increase in one is balanced by a decrease in the other.

3 MODELING

Most quantitative models concern uncertainty. An exception is the fuzzy sets theory that addresses imprecision.

3.1 The quantification of imprecision: fuzzy sets

Imprecision is essentially represented by a disjunctive information that characterizes a set of possible values to which the actual value is known to belong. Recently the classical concept of set has been extended to fuzzy sets that have been used to characterize 'ordered' disjunctive information.

Classically, sets are crisp in the sense that one element either belongs to a set or is excluded from it. Zadeh (1965) introduces the idea of non-crisp sets, called fuzzy sets. Fuzziness is a property related to the use of vague predicates like in 'John is tall'.

The predicates are vague, fuzzy because the words used to define them are themselves ill defined, vague, fuzzy . The idea is that belonging to a set admits a degree that is not necessarily just 0 or 1 as is the case in the classical set theory. For some elements of the universe of discourse, one cannot say that they belong or not to the set. At most one can assess some degree of membership $\mu_A(x)$ of the element x in the fuzzy set A. This function generalizes the classical indicator function $I_A(x)$ of a set:

$$I_A(x) = 1 \text{ if } x \in A \qquad I_A(x) = 0 \text{ if } x \notin A$$

Zadeh replaces the range {0, 1} by the interval [0, 1].

New concepts like fuzzy numbers (e.g. several, few), fuzzy probability (likely), fuzzy quantifiers (most), fuzzy predicates (tall), and the impact of linguistic hedges (very) can be formalized (Dubois and Prade, 1980). Classical set operators like union, intersection and negation have been generalized. The most classical solution is based on the min-max operators:

$$\mu_{\overline{A}}(x) = 1 - \mu_A(x)$$

$$\mu_{A \cup B}(x) = \max(\mu_A(x), \mu_B(x))$$

$$\mu_{A \cap B}(x) = \min(\mu_A(x), \mu_B(x))$$

The law of excluded middle does not apply to fuzzy sets. Indeed $\mu_{A \cap \overline{A}}(x) = \min(\mu_A(x), \mu_{\overline{A}}(x))$ can be larger than 0. This must look odd at first sight. This translates nothing but the fact that one can be somehow tall and not tall simultaneously, a perfectly valid property.

Mathematically, the fuzzy sets theory generalizes the concept of set. The model can be used wherever sets can be used and therefore is not restricted to any particular form of imperfect data. Its simplest domain of application is the modeling of imprecision and vagueness. Fuzziness creates an order among the possible values in which the actual value is known to belong.

Several authors have tried to disregard the fuzzy sets theory by claiming that it is subsumed by the probability measure. This attitude completely misfires.

The fuzzy set theory concerns the belonging of a well-defined individual to an ill-defined set, whereas probability concerns the belonging of a not yet defined individual to a well-defined set. Introducing random sets does not change the conceptual picture. Of course there are mathematical relations between the two theories, but the problem is not with the mathematical comparison but with a comparison of the problems they try to model. Fuzziness deals with imprecision, probability with uncertainty.

Of course, fuzziness induces uncertainty. One could defend that when I know that John is tall, I can build a probability measure on John's height. This does not mean that the grade of membership is a probability (Smets 1985)

3.2 Possibility measure

Incomplete information such as "John's height is above 170" implies that any height h above 170 is possible and any height equal or below 170 is impossible. This can be represented by a 'possibility' measure defined on the height domain whose value is 0 if $h < 170$ and 1 if h is ≥ 170 (with 0 = impossible and 1 = possible). Ignorance results from the lack of precision, of specificity of the information "above 170".

When the predicate is vague like in "John is tall', possibility can admit degrees, the larger the degree, the larger the possibility. But even though possibility is often associated with fuzziness, the fact that non fuzzy (crisp) events can admit different degrees of possibility is shown in the following example. Suppose there is a box in which you try to squeeze soft balls. You can say: it is possible to put 20 balls in it, impossible to put 30 balls, quite possible to put 24 balls, but not so possible to put 26 balls...These degrees of possibility are degrees of realizability and totally unrelated to any supposedly underlying random process.

Let $\Pi : 2^{\Omega} \rightarrow [0, 1]$ be the possibility measure defined on a space Ω with $\Pi(A)$ for $A \subseteq \Omega$ being the degree of possibility that A (is true, occurs...). The fundamental axiom is that the possibility $\Pi(A \vee B)$ of the disjunction of two propositions A and B is the maximum of the possibility of the individual propositions $\Pi(A)$ and $\Pi(B)$. (Zadeh 1978, Dubois and Prade, 1985):

$$\Pi(A \vee B) = \max (\Pi(A) , \Pi(B)). \qquad (3.1)$$

Let Ω be the universe of discourse on which a possibility measure Π is defined. Related to the possibility measure $\Pi : 2^{\Omega} \rightarrow [0, 1]$, one can define a possibility distribution $\pi : \Omega \rightarrow [0, 1]$,

$$\pi(x) = \Pi(\{x\}) \qquad \text{for all } x \in \Omega.$$

Thanks to (3.1), one has

$$\Pi(A) = \max_{x \in A} \pi(x) \qquad \text{for all A in } \Omega.$$

It generalizes into :

$$\Pi_A(X) = \max_{x \in \Omega} \min(\pi_A(x), \mu_X(x)),$$

where Ω is the domain of x and X is a fuzzy subset of Ω. One can thus express the possibility that the height of a person is about 180 cm. given the person is tall.

A very important point in possibility theory (and in fuzzy set theory) when only the max and min operators are used is the fact that the values given to the possibility measure or to the grade of membership are not intrinsically essential. The only important element of the measure is the order they create among the elements of the domain. Indeed the orders are invariant under any strictly monotonous transformation. Therefore a change of scale will not affect conclusions.

This property explains why authors insist on the fact that possibility theory is essentially an ordinal theory, a nice property in general. This robustness property does not apply once addition and multiplication are introduced as is the case with probability and belief functions.

3.3 Relation between fuzziness and possibility

Zadeh has introduced both the concept of fuzzy set (1965) and the concept of possibility measure (1978). The first allows one to describe the grade of membership of a well-known individual in an ill-defined set. The second allows one to describe the individuals who satisfy some ill-defined constraints or who belong to some ill-defined sets.

For instance $\mu_{Tall}(h)$ quantifies the membership of a person with height h in the set of *Tall* men and $\pi_{Tall}(h)$ quantifies the possibility that the height of a person is h given that the person belongs to the set of *Tall* men. Zadeh's possibilistic principle postulates the following equality :

$$\pi_{Tall}(h) = \mu_{Tall}(h) \qquad \text{for all } h \in H$$

where H is the set of height = $[0, \infty)$

The writing is often confusing and would have been better written as

$$\text{If } \mu(Tall|h) = x \text{ then } \pi(h|Tall) = x \qquad \text{for all } h \in H$$

The last expression avoids the confusion between possibility and fuzziness. It shows that they share the same scale without implying that a possibility is a membership and *vice versa*. The last expression clearly indicates the domain of the measure (sets for the grade of membership μ and height for the possibility distribution π) and the background knowledge (the height for μ and the set for π).

3.4 Probability theory

Probability measure quantifies the degree of probability P(A) (whatever probability means) that an arbitrary element $X \in \Omega$ belongs to a well-defined subset $A \subseteq \Omega$ It satisfies the following property :

P1: $P(\emptyset) = 0 \qquad P(\Omega) = 1$

P2: For all $A,B \subseteq \Omega$, if $A \cap B = \emptyset$, $P(A \cup B) = P(A) + P(B)$

P3: For all $A, B \subseteq \Omega$, if $P(B)>0$, then $P(A|B) = P(A \cap B) / P(B)$

where P(A|B) is the probability of that $X \in A$ given it is known that $X \in B$. Such a definition can be extended to fuzzy events (Zadeh, 1968, Smets 1982) which further enhances, if still needed, the difference between probability and fuzziness. As an example consider the probability that the next man who enters the room is a tall man. Could we say that such a probability is .7, or is that probability itself a fuzzy probability? This is still unresolved and might explain today's lack of interest in that concept.

3.5 Credibility: the transferable belief model

Information can induce some subjective, personal credibility (hereafter called belief) that a proposition is true. There is no necessity for beliefs at the credal states (the psychological level where beliefs are entertained) to be quantified by probability measures. The coherence principle advanced by the Bayesians to justify probability measures is adequate in a context of decision, but it cannot be used when all one wants to describe is a cognitive process.

Beliefs can be entertained outside any decision context. In the transferable belief model (Smets 1988, Smets and Kennes, 1994) we assume that beliefs at the credal level are quantified by belief functions (Shafer 1976). When decisions must be made, the belief held at the credal level induces a probability measure held at the so-called 'pignistic' level (the level at which decisions are made).

This probability measure will be used in order to make decisions using the expected utilities theory. Relations between belief functions held at the credal level and probabilities held at the "pignistic" level are given in Smets (1990).

4 Combining Models of Ignorance

The various forms of ignorance can be encountered simultaneously, and it is necessary to be able to integrate them.

The first problem is to investigate the connections between the probability theory in its frequency approach and the physical possibility theory. The next problem is to investigate the connections between subjective probability measures, belief functions and epistemic possibility measures. Finally, one must establish the connections between the physical properties and the epistemic properties. There is further the problem of extending all these theories when the propositions involved are fuzzy.

Almost no work has been done in this area. However, its importance for data fusion is obvious: when several sensors provide information, how do we recognize the nature of the ignorance involved and select the appropriate model; how do we collapse them into more compact forms; how do we combine them; how do we take into consideration the redundancies, the correlations and the contradictions? All these problems must be studied and the implementation of potential solutions tested.

CONCLUSION

The conclusions are more in the form of a plea for ecumenism. We have shown that imprecision and uncertainty are really multiform, and that none of the models available today can fit with all forms of imperfect data. The real problem is to have an open mind attitude and to avoid the dogmatic attitude that leads to claims like: 'I can do *everything* with *my* theory' or 'the forms not covered by *their* models are useless'. We will not insist on the arrogance that underlies such claims.

Both forcing fuzzy sets into the probability mold and claiming that the fuzzy sets theory embeds all other theories are usually unfounded. Each model aims at describing some forms of imperfection. They are complementary, not concurrent.

When confronted with imperfect data, the user should first try to realize the form of imperfection he/she is facing, then see which model is the most appropriate. The real challenge is in recognizing the nature of the imprecision and uncertainty encountered in a given problem. This paper tried to give some hints in that direction.

KEYWORDS

Uncertainty, imprecision, fuzzy sets theory, possibility theory, probability theory, transferable belief model.

REFERENCES

[1] Dubois D., Prade H., Fuzzy sets and systems: theory and applications, Academic Press, New York (1980).

[2] Dubois D., Prade H., Théorie des possibilités, Masson, Paris.(1985) [English Version: Theory of Possibility, Plenum, London (1988)

[3] Shafer G., A mathematical theory of evidence, Princeton Univ. Press, Princeton, NJ (1976).

[4] Smets P., Probability of a Fuzzy Event : an Axiomatic Approach, Fuzzy Sets and systems, 7, 153-164 (1982).

[5] Smets P., Probability of a Fuzzy Event, In : Systems and Control Encyclopedia, Singh M. G. (ed) Pergamon, Oxford, 1802-1805 (1985).

[6] Smets P., Belief functions, in Non standard logics for automated reasoning, Smets P., Mamdani A., Dubois D. , Prade H. (eds), Academic Press, London, pp. 253-286.(1988)

[7] Smets P., Constructing the pignistic probability function in a context of uncertainty. Uncertainty in Artificial Intelligence 5, Henrion M., Shachter R.D., Kanal L.N., Lemmer J.F. (eds), North Holland, Amsterdam, , 29-40 (1990).

[8] Smets P., Varieties of ignorance. Information Sciences, 57-58:135-144 (1991).

[9] Smets P., Kennes R., The transferable belief model. Artificial Intelligence 66:191-234 (1994).

[10] Smithson M., Ignorance and Uncertainty: Emerging Paradigms, Springer-Verlag, New York (1989).

[11] Zadeh L.A., Fuzzy sets, Inform.Control, 8:338-353 (1965).

[12] Zadeh L.A., Probability measures of fuzzy events, J. Math. Anal. Appl., 23: 421-427 (1968).

[13] Zadeh L.A, Fuzzy sets as a basis for a theory of possibility, Fuzzy Sets and Systems, 1:3-28, (1978).

ILL-DEFINED DATA IN INFORMATION SYSTEMS : FROM FUZZY PATTERN MATCHING TO DATA FUSION

Didier DUBOIS and Henri PRADE

Institut de Recherche en Informatique de Toulouse, Université Paul Sabatier,
118 route de Narbonne, 31062 Toulouse Cedex, France

ABSTRACT

Data are often pervaded with imprecision or with uncertainty. Such imperfect pieces of information raise several theoretical problems if we want to use them in information systems : representation of their contents, evaluation of their relevance with respect to a query, combination of such pieces of information provided by different sources. The paper surveys what is offered by the fuzzy set and possibility theory for dealing with these three problems.

RESUME

Les données sont souvent imprécises ou incertaines. Le traitement de telles données soulève le problème de leur représentation, de l'évaluation de leur pertinence par rapport à une requête, et de la combinaison de telles informations quand elles proviennent de sources différentes. Cet article présente une brève vue d'ensemble de ce qu'offre la théorie des possibilités pour traiter ces trois problèmes.

1 INTRODUCTION

By a piece of data, we mean here a subset of values attached to a variable representing an attribute value of an object of interest. This subset S may i) reduce to a singleton of the attribute domain D (e.g., "John's height is 1.82 meter", S ={1.82}), and then the datum is precise and certain ; ii) be an ordinary subset (larger than a singleton) of the attribute domain, (e.g, "John's height is between 1.80 and 1.85 meter", i.e., S = [1.80, 1.85]) and then the piece of information is *imprecise* ; iii) be a fuzzy subset of the attribute domain with a restricted support, i.e., support(S) = {s, $\mu_S(s)>0$} is a strict subset of the domain D. This means that the values outside support(S) are *certainly impossible* for the variable underlying the considered attribute, while μ_S rank-orders the values according to their plausibility on support(S). We can then represent vague statements, like "John is tall", (i.e. S = tall) in a given context ; iv) be a fuzzy set with an unrestricted support ; i.e. absolutely no value of the attribute domain can be discarded, even if some values are more plausible than others. The piece of information thus represented is said to be *uncertain*. Note that a datum may be precise but uncertain. This is so if only *one* value x_0 has a maximum plausibility level while all the others have a uniform (much) smaller one α. Then our certainty that x_0 is the genuine value is given by $1- \alpha$ in possibility theory (the more plausible a value distinct from x_0, the less certain x_0 is the genuine value). The (fuzzy) subsets used for representing imprecise or uncertain data play the role of a *possibility distribution* which restricts the more or less plausible/possible values of the considered variable. Once stored in a data base, imprecise or uncertain data relevant for a query must be retrieved by pattern matching, discussed in Part I of the paper. In Part II we deal with the fusion of imprecise or uncertain data provided by different sources of information, always in the framework of the possibility theory. This paper briefly presents the authors' approach to data management pervaded with imprecision and uncertainty. More details on fuzzy pattern matching are found in [1], [3], [7] and on multiple source information combination in [4], [6], [8], [9].

2 Fuzzy Pattern Matching

Basically, a pattern matching process involves a pattern describing some requirements and a data base where items or pieces of information about objects, are gathered. In the simplest case, data identical to the specification expressed in the pattern are looked for and retrieved. When data are pervaded with imprecision and uncertainty or when the pattern includes vague specifications, matching between the pattern and an item is naturally a matter of degree. A fuzzy pattern matching technique has been developed [1], [3] based on [0,1]-valued possibility [11] and necessity measures, in order to estimate the compatibility between an item and what is required by the pattern. In this approach each component of an item and each elementary requirement are respectively associated with a possibility distribution and a fuzzy set. A possibility distribution is a convenient tool for representing imprecise, uncertain or vague pieces of data pertaining to single-valued attributes, while a fuzzy set models a soft constraint whose satisfaction is a matter of degree.

The basic dissymmetry of the pattern-data matching is preserved by this modeling convention. Indeed, a fuzzy pattern represents an ill-bounded class of objects, while a fuzzy item represents an ill-known object whose precise description is not available. Namely let P and D be respectively a pattern component and an item component pertaining to the same single-valued attribute, which are to be compared. P and D refer to the same scale U conveying their meanings. Let μ_P be the membership function associated with atom P and π_D the possibility distribution attached to D. Both are mappings from U to [0,1]. Let u be an element of U. Then $\mu_P(u)$ is the grade of compatibility between the value u and the meaning of P. Namely $\mu_P(u) = 1$ means total compatibility with P and $\mu_P(u) = 0$ means total incompatibility with P. By contrast, $\pi_D(u)$ is the grade of possibility that u is the (unique) value of the attribute describing the object modeled by the item. D is a fuzzy set of *possible* values (only one of which is the genuine value of the ill-known attribute), while P is a fuzzy set of *more or less* compatible values. For instance $\pi_D(u) = 1$ means that u is totally possible, while $\pi_D(u) = 0$ means that u is totally impossible as an attribute value of the object to which the item pertains. In the following μ_P and π_D are always supposed to be normalized, i.e., there is always a value which is totally compatible with P, and a value totally possible in the range D.

Two scalar measures are used to estimate the compatibility between a pattern atom P and its counterpart D in the item list, i.e. a degree of possibility of matching $\Pi(P ; D)$ and a degree of necessity of matching $N(P ; D)$ respectively defined [11], [3] by

$$\Pi(P ; D) = \sup_{u \in U} \min(\mu_P(u), \pi_D(u)), \qquad N(P ; D) = \inf_{u \in U} \max(\mu_P(u), 1 - \pi_D(u)).$$

$\Pi(P ; D) \geq N(P ; D)$ always holds if D is normalized ($\exists u, \pi_D(u)=1$).

The limiting cases where $\Pi(P ; D)$ and $N(P ; D)$ take values 0 and 1 are useful to study in order to lay bare the semantics of these indices. For any fuzzy set, F on U, let $F^\circ = \{u \in U \mid \mu_F(u) = 1\}$ be the core of F, and $s(F) = \{u \in U, \mu_F(u) > 0\}$ its support. Then it can be checked that

(i) $\Pi(P ; D) = 0$ iff $s(P) \cap s(D) = \emptyset$, (ii) $\Pi(P ; D) = 1$ iff $P^\circ \cap D^\circ \neq \emptyset$,
(iii) $N(P ; D) = 1$ iff $s(D) \subseteq P^\circ$, (iv) $N(P ; D) > 0$ iff $D^\circ \subset s(P)$ (strict inclusion).

The measure $\Pi(P ; D)$ estimates to what extent P and D can refer to the same value u; i.e. $\Pi(P ; D)$ is a degree of overlapping of the fuzzy set of values compatible with P, with the fuzzy set of possible values of D. The measure $N(P ; D)$ estimates to what extent it is necessary (i.e., certain) that the value to which D refers is among those compatible with P; i.e. $N(P ; D)$ is a degree of inclusion of the set of possible values of D into the set of values compatible with P. However it is based on a stronger view of inclusion between fuzzy sets than the usual one ($\pi_D \leq \mu_P$); the latter only implies $N(P ; D) \geq 0.5$, while here, $N(P ; D) = 1$ if and only if all the more or less possible values restricted by π_D are compatible with P to degree 1. Besides, it is worth noticing that $N(F ; F) = 1$ if and only if μ_F is the membership function of an ordinary subset of U which stands for μ_P and π_D. Indeed, in this case we are completely certain that an ill-known value which is only known to belong to D = F satisfies the requirement, expressed by the pattern P, of belonging to F, whatever its precise value. Otherwise we only have $N(F ; F) \geq 1/2$. Indeed when P = D have a fuzzy meaning, we cannot be sure that the degree of matching $\mu_P(x)$ of the component item x fuzzily restricted by $\pi_D = \mu_P$, is high. Note that when D is precise, i.e., $\exists d, \pi_D(d) = 1$, and $\forall u \neq d, \pi_D(u) = 0$ which can be written $D = \{d\}$ then $\Pi(P ; \{d\}) = N(P ; \{d\}) = \mu_P(d)$. The table below summarizes the respective locations of $\Pi(P ; D)$ and $N(P ; D)$ in [0,1] when P and/or D is precise, imprecise (but clear-cut) or fuzzy.

	Precise	Imprecise	
P	D = {d}	D (non-fuzzy)	D Fuzzy
Precise P = {p}	$\Pi = N = \begin{cases} 1 \text{ if } p = d \\ 0 \text{ if } p \neq d \end{cases}$	$N = 0$ $\Pi = \begin{cases} 1 \text{ if } p \in D \\ 0 \text{ otherwise} \end{cases}$	$N < 1$
Imprecise (non-fuzzy)	$\Pi = N = \begin{cases} 1 \text{ if } d \in P \\ 0 \text{ if } d \notin P \end{cases}$	$N = \Pi = \begin{cases} 1 \text{ if } D \subseteq P \\ 0 \text{ if } D \cap P = \emptyset \end{cases}$	$N > 0 \Rightarrow \Pi = 1$
Fuzzy	$\Pi = N = \mu_P(d)$	$\Pi \geq N$	$\Pi \geq N$

The elementary measures of possibility and necessity are aggregated separately in order to obtain two global measures between the whole pattern and the whole item. When the pattern expresses a conjunction of elementary requirements "P_1 and... P_n", this aggregation is performed using the 'min' operation and preserves the respective semantics of the measures in terms of possibility and necessity. We have e.g. see [4]:

$$\Pi(P_1 \times \ldots \times P_n ; D_1 \times \ldots \times D_n) = \min_{i=1,\ldots,n} \Pi(P_i ; D_i),$$
$$N(P_1 \times \ldots \times P_n ; D_1 \times \ldots \times D_n) = \min_{i=1,\ldots,n} N(P_i ; D_i),$$

where P_i and D_i are supposed to be defined on the same domain U_i, and where $\times$ denotes the Cartesian product defined for two fuzzy sets F_i and F_j by $\mu_{F_i \times F_j}(u_i,u_j) = \min(\mu_{F_i}(u_i), \mu_{F_j}(u_j))$. The attributes appearing in the pattern and the item are supposed to be non-interactive [11]. This means that a restriction on the possible values of item D_i expressed by π_{D_i} has no influence on the possibility distribution π_{D_j} restricting the possible values of the item D_j; no constraint links the values of D_i and D_j. The definition of F_i x F_j supposes that all the parts of the pattern (which expresses what is required) are equally important from the user's point of view. Disjunctive requirements are dealt with changing min into max, and $P_1 \times \ldots \times P_n$ into $P_1 + \ldots + P_n$ with $\mu_{P_i+P_j} = \max(\mu_{P_i}, \mu_{P_j})$.

Usually a single pattern expressing a requirement is matched against a whole database. A user wants to get the data that yield the best compatibility measures with respect to the pattern. Thus, we are faced with the problem of ranking the items, each of them characterized by a pair of degrees. In order to establish a classification of several items with respect to a pattern, we may consider the following remarks:

- The necessity degree is of greater importance than the possibility one, because when positive it is nearly certain that the item matches the requirement. It is reasonable to rank-order the items in terms of N(P ; D) and to use Π(P ; D) in order to break ties.
- If we obtain a necessity degree equal to zero and a possibility degree equal to 1 for each item, it means that what is known about the data is too imprecise or vague with respect to the pattern. We may then try to improve the knowledge about the data.
- The more precise D, the closer to zero Π(P ; D) – N(P ; D) is. However the converse is false.

•Generally, the Pareto ordering is used to rank pairs of degrees, as follows: (Π,N) is greater than (Π',N') if and only if Π > Π' and N ≥ N', or Π ≥ Π' and N > N'. However, there are situations where Π(P ; D) > Π(P ; D') and N(P ; D) < N(P ; D'). Here P and D stand for a compound pattern or item. The Pareto ordering is only partial. Thus we may obtain *several* data which somewhat best match the pattern. We note that if D and D' are maximal elements for the Pareto ordering, N(P ; D) < N(P ; D') < Π(P ; D') < Π(P ; D) indicating that D' is the best item from the cetainty point of view.

3 DATA FUSION

The problem of combining pieces of evidence from several sources of information is encountered in various applications, particularly in i) sensor fusion, i.e., when pieces of information from different sensors are to be aggregated, ii) multiple source interrogation systems where each of the sources can provide precise, imprecise or uncertain information about values of interest, iii) expert opinion pooling, when different individual statements must be synthesized. Our basic claim is that there can be no unique mode of combination, satisfactory in any situation, even when the framework for representing information is chosen.

3.1 Some limitations of the probabilistic approach

The probabilistic approach can be criticized for several reasons ; see [6] for details:

•First, identifying a probability distribution requires more information than an expert can supply. Many distribution functions correspond to given .5 and .95 fractiles and prescribed modes. Experts prefer to supply intervals rather than point-values because their knowledge is not only of limited reliability but also tainted with imprecision.
•The consensus method has a basic flaw in the context of reliability: it is a voting-like procedure. Indeed if two reliable experts have conflicting opinionsabout the value of x, so that one gives a low value to x, and the other givesa high value to x, the consensus method will supply a probability distribution whose mean value is medium, i.e., a value on which both experts agree as not being the true one.
•Another interpretation of the consensus method amounts to considering the expert data as issued from a single random source. The underlying homogeneity assumption is questionable in the case of experts some of whom are wrong, or multiple sources, some of which are erratic or erroneous.
•The main drawback of the Bayesian method seems to be, as usual, the need for a priori knowledge about the value of x. In other words the analyst who looks for expert advice must be an expert himself. However in many cases the analyst has no idea about the value of x and all that he may learn is about the reliability of experts, by a technique like the one by Cooke [2], that inspired the one described in [9].

The possibilistic approach outlined here tries to cope with most of the difficulties faced by the probabilistic approach on pooling imprecise information from heterogeneous sources; its main features are: faithfulness in representing subjective data, no need for a priori knowledge, and a variety of pooling methods whose choice depends on the reliability of sources and the level of conflict between their opinions.

3.2 A logical view of the combination of uncertainty

The general ideas of the possibilistic approach to the aggregation of information from several distinct sources are 1) there is no unique combination mode; 2) the choice of the combination mode depends on an assumption about the reliability of sources, as formulated by an analyst. No a priori knowledge of the inquired parameter is needed, and the sources, viewed in parallel, must be combined symmetrically if all sources are equally reliable. There are basically two extreme modes of symmetric combination: the conjunctive mode when all sources agree and are considered as reliable; and the disjunctive mode when sources disagree so that at least one of them is wrong.

Since a particular case of possibility distribution is the characteristic function of a set, the basic combination modes in the possibilistic setting are conjunctions and disjunctions. There exist several possible choices among pointwise fuzzy set-theoretic operations for defining conjunctive and disjunctive combinations of possibility distributions. Let π_i be the possibility distribution supplied by source i, i = 1,n. Define

$$\forall s \in S, \pi_{\wedge}(s) = *_{i=1,n}\, \pi_i(s) \text{ (fuzzy set intersection)} ;\ \pi_{\vee}(s) = \perp_{i=1,n}\, \pi_i(s) \text{ (fuzzy set union)}$$

where $*$ and $\perp$ are [0,1]-valued operations defined on $[0,1] \times [0,1]$ which enjoy the duality relation $a \perp b = 1 - (1 - a) * (1 - b)$ in order to express De Morgan's law. Candidates for $*$ and $\perp$ are so-called triangular norms and co-norms; i.e., $*$ is an operation which is associative, commutative, increasing (in the wide sense) in both places and such that $a * 1 = a$ and $0 * 0 = 0$. The main continuous solutions are $a * b = \min(a,b)$, $a * b = a \cdot b$ (product) and $a * b = \max(0, a + b - 1)$, which leads to $a \perp b = \max(a,b)$, $a \perp b = a + b - a \cdot b$, and $a \perp b = \min(1, a + b)$; see [3] for details.

As already said, the conjunctive aggregation makes sense if all the sources are considered as equally and fully reliable, while the disjunctive aggregation corresponds to a weaker reliability hypothesis, namely, in the group of sources there is at least one reliable source for sure, but we do not know which one. In the conjunctive case, the min operation corresponds to a purely logical view of the combination process: the source which assigns the least possibility degree to a given value is considered as the best-informed with respect to this value. Note that with min, when all sources perfectly agree ($\forall\ i \neq j$, $\pi_i = \pi_j$), there is no reinforcement effect. Generally, agreement between experts is due to common background, and the idempotence of min. deals with such a kind of redundancy. On the contrary if $*$ = product, if all the sources agree that a value s is not fully possible, this value will receive a possibility degree strictly smaller than $\min_{i=1,n} \pi_i(s)$, i.e., the lack of complete possibility is reinforced; a necessary condition for choosing such an operation is the independence of the sources. This assumption may be more adapted to sensor fusion problems.

An important issue with conjunctive combination is the fact that the result may be subnormalized, i.e., it may happen that $\nexists s, \pi_\wedge(s) = 1$. In that case it expresses a conflict between the sources. Clearly the conjunctive mode makes sense if all the π_i significantly overlap, i.e., $\exists s, \forall i, \pi_i(s) = 1$, expressing that there is at least a value of s that all sources consider as completely possible. If $\forall s, \pi_\wedge(s)$ is significantly smaller than 1 this mode of combination is debatable since in that case at least one of the sources or experts is likely to be wrong, and a disjunctive combination might be more advisable. In case of subnormal results ($\sup \pi_\wedge < 1$), we may think of renormalizing $\pi_\wedge$, thus leading to the new rule (here for n = 2)

$$\forall s \in S, \pi(s) = \pi_\wedge(s) / h(\pi_1, \pi_2) \qquad (1)$$

where $\pi_\wedge(s) = \pi_1(s) * \pi_2(s)$ and $h(\pi_1,\pi_2)$ is the height of the intersection of π_1 and π_2 defined by $h(\pi_1,\pi_2) = \sup_{s \in S} \pi_1(s) * \pi_2(s) = \sup \pi_\wedge$ where $h(\pi_1,\pi_2)$ is a natural measure of overlap between two possibility distributions.

3.3 Towards adaptive rules of combination

Clearly the renormalization erases the conflict between the sources as expressed by the subnormalization. Even if it is good that the result of the combination focuses on the values on which all the sources partially agree (in the sense that none of them gave to these values a possibility equal to 0), it would be better that the result also keep track of the conflict in some way. However, the normalized rule may be very sensitive to rather small variations of possibility degrees around 0; indeed the rule is not continuous in the vicinity of the total conflict expressed by $h(\pi_1,\pi_2) = 0$. A natural idea for keeping track of a partial conflict is to discount the result given by (1) by a weight corresponding to the lack of normalization, i.e., $1 - h(\pi_1,\pi_2)$. Namely $1 - h(\pi_1,\pi_2)$ is viewed as the degree of possibility that both sources are wrong since when $h(\pi_1,\pi_2) = 1$,the two sources are supposed to be right. As an example of such a discounted conjunctive combination rule we have

$$\forall s \in S, \pi'(s) = \max(\pi(s), 1 - h(\pi_1, \pi_2)) = \max((\pi_1(s)*\pi_2(s))/h(\pi_1, \pi_2)), 1-h(\pi_1, \pi_2)).$$

The amount of conflict $1 - h(\pi_1,\pi_2)$ induces a uniform level of possibility for all values outside the ones emerging in the subnormalized intersection of π_1 and π_2, i.e., the result of the combination is not fully certain. Clearly, $\lim_{h(\pi'_1,\pi_2)\to 0} \pi'(s) = 1$, so that the discontinuity effect is coped with, resulting in total ignorance.

When $h(\pi_1,\pi_2)$ is low, we might be more optimistic and assume that the discrepancy between the sources is due to one of them being wrong, not both. Then, instead of transferring the amount of conflict as a uniform degree of possibility on the whole referential S, we may limit ourselves to the union of F_1 and F_2 ($\pi_1 = \mu_{F_1}, \pi_2 = \mu_{F_2}$), considering that there is no reason to put a non-zero possibility degree on values that both sources agree to consider as impossible. Hence the more elaborated adaptive rule [5],[6]

$$\forall s \in S, \pi_{AD}(s) = \max(\min(\pi_1(s),\pi_2(s)) / h(\pi_1, \pi_2), \min(\max(\pi_1(s),\pi_2(s)), 1 - h(\pi_1, \pi_2)))$$

$\pi_{AD}(s)$ evaluates at x = s the fuzzy logic expression: "the two sources are reliable and claim that x = s, OR (they are not both reliable AND one of them is reliable and claims that x = s)", where $h(\pi_1,\pi_2)$ evaluates to what extent both sources are reliable, AND and OR being expressed by min and max.

As pointed out in [6], the fuzzy set R of more or less reliable experts obtained from the assessment procedure is useful to partition the set of experts into classes K_1, $K_2,\ldots, K_q$ of equally reliable ones, where K_j corresponds to a higher reliability level than K_{j+1}, for j = 1,q. Then the above symmetric aggregation schemes extended to n sources can be applied to each class K_j. The combination between results obtained from the K_j's can be performed according to the following principle: the response of group K_2 is used to refine the response of group K_1 insofar as it is consistent with it. If π_1 is obtained from K_1 and π_2 from K_2, the degree of consistency of π_1 and π_2 is again $h(\pi_1,\pi_2) = \sup_s \min(\pi_1(s),\pi_2(s))$ and the following combination rule has been proposed (see, e.g., [5], [10]): $\pi^{\widehat{1\text{-}2}} = \min(\pi_1, \max(\pi_2, 1 - h(\pi_1, \pi_2))$. When $h(\pi_1, \pi_2) = 0$, K_2 contradicts K_1 and only the opinion of K_1 is retained, i.e., $\pi^{\widehat{1\text{-}2}} = \pi_1$. See [6] for more details.

KEYWORDS

Pattern matching, uncertainty, imprecision, possibility theory, adaptative combination, data fusion, expert opinion pooling.

REFERENCES

[1] Cayrol M., Farreny H., Prade H., Fuzzy pattern matching, Kybernetes, 11, 103-116 (1982).
[2] Cooke R.M., Experts in Uncertainty, Oxford University Press, Oxford, UK (1991).
[3] Dubois D., Prade H., Possibility Theory — Plenum Press, New York (1988).
[4] Dubois D., Prade H., On the relevance of non-standard theories of uncertainty in modeling and pooling expert opinions, Reliability Eng. & Syst. Safety, 36, 95-107 (1992).
[5] Dubois D., Prade H., Combination of fuzzy information in the framework of possibility theory, In: Data Fusion in Robotics and Machine Intelligence, Abidi M.A., Gonzalez R.C. (eds.) 481-505, Academic Press, New York (1992).
[6] Dubois D., Prade H., Possibility theory and data fusion in poorly informed environments, Control Engineering Practice, 2, (5), 811-823 (1994).
[7] Dubois D., Prade H., Testemale C., Weighted fuzzy pattern matching, Fuzzy Sets and Systems, 28, 313-331 (1988).
[8] Sandri S., Besi , Dubois D., Mancini G., Prade H., Testemale C., Data fusion problems in an intelligent data base interface. In: Reliability Data Collection and Use in Risk and Availability Assessment, Colombari V. (ed.) 655-670, Springer Verlag, Berlin (1989).
[9] Sandri S., Dubois D., Kalfsbeek H., Elicitation, assessment and pooling of expert judgement using possibility theory, IEEE Trans. on Fuzzy Systems, 3(3), 313-335 (1995)
[10] Yager R., Non-monotonic set-theoretic operators, Fuzzy Sets & Syst., 42, 173-190 (1991).
[11] Zadeh L., Fuzzy sets as a basis for a theory of possibility, Fuzzy Sets & Syst., 1, 3-28 (1978).

Error and Uncertainty in Physical Measurements

E. Richard COHEN

Science Center Rockwell International Corporation, 1049 Camino dos Rios, Thousand Oaks, CA 91360, USA

Abstract

'Error' and 'uncertainty' are two complementary, but distinct, aspects of the characterization of measurements. 'Error' is the difference between a measurement result and the value of the measurand while 'uncertainty' describes the reliability of the assertion that the stated measurement result represents the value of the measurand. The analysis of error considers the variability of the results when the measurement process is repeated. The evaluation of uncertainty considers the observed data to be given quantities from which the estimates of certain parameters (the measurement results) are to be deduced. The failure to distinguish between these two concepts has led to inconsistency, and a lack of uniformity in the way uncertainties have been expressed. The 1993 ISO (International Organization for Standardization) Guide to the Expression of Uncertainty in Measurements *is the first international attempt to establish this uniformity and makes no distinction in the treatment of contributions to the total uncertainty in a measurement result between those arising from "random errors" and those arising from "systematic errors."*

Resume

L'erreur et l'incertitude sont deux aspects complémentaires, mais distincts, de la caractérisation des mesures. L' "erreur" est la différence entre la mesure résultante et la valeur du mesuré alors que l' "incertitude" correspond au bien fondé de l'assertion que le résultat énoncé de la mesure représente bien la valeur de la chose mesurée. L'analyse d'erreur considère la variabilité des résultats lorsque les mesures sont répétées. Pour l'évaluation de l'incertitude, on associe à une donnée observée certaines valeurs dont on soustrait les estimations de certains paramètres (résultat des mesures). Ne pas distinguer clairement ces deux concepts a conduit à une incohérence et à une absence d'uniformité dans l'utilisation des incertitudes. Le "Guide de l'expression de l'incertitude dans les mesures" publié en 1993 par l'ISO n'introduit aucune distinction dans le traitement des contributions à l'incertitude totale provenant d'"erreurs aléatoires" et celles issues d'erreurs systématiques.

1 INTRODUCTION

In order to analyze the variability in the results of a set of measurements, it must be clear what that variability represents. If, for example, the measurements are the heights of the incoming freshmen class at the university, one expects a variation from one individual to another. Standard statistical techniques may be used to estimate the mean, dispersion, and other characteristics of these measurements, and, for example compare them with the results from previous years. In this case the precision of the measurements themselves is not a primary concern; the observations will probably be rounded to the nearest centimeter, thereby washing out most of the error of an individual measurement. Measurements such as these, in which the objective was the characterization of a population of items, provided the basis for the language and the development of statistics starting approximately a century ago that can be said to have culminated in the work of R. A. Fisher and his school.

Although the same language continues to be applied, there is a difference in concepts when statistical analysis is applied to measurements that are observations of the value of a specific physical quantity, and the purpose is to describe the errors in those measurements. In the traditional (or 'orthodox') frequentist approach the fiction is established that there is a population of errors waiting to be observed, and that those errors can be sampled in the same way that individuals are sampled. At the same time that the nurse at Student Health determines their heights, let the students measure the height of the nurse. Both sets of data, measurements of the students' height by the nurse and the measurements of the nurse's height by the students, will show a range of observed values, but these variations are not of the same kind. In the first set the variations arise primarily from differences in the measured quantities, in the second set the variations arise from differences in the observation of one (presumably unique) quantity. The data on the nurse's height can be used to determine an estimate of the height of the nurse, and they can also be used to evaluate the ability of students to make measurements. These are obviously different objectives, and the language and algorithms that are used for one need not necessarily be appropriate for the other.

2 VOCABULARY

Before one can discuss the ideas of 'error' and 'uncertainty' it will be necessary to define the concepts and establish a notation and a vocabulary. In 1993 ISO (the International Organization for Standardization) published two documents, the *Vocabulaire international des termes fondamentaux et généraux de métrologie* (VIM)[1] and the *Guide to the Expression of Uncertainty in Measurement* (*Guide*)[2]. These documents provide the basis for the discussion here.

2.1 Physical quantity

A physical quantity and its value will be denoted by a capital letter (Y) and observations of the quantity will be denoted by a lower case letter (y). An estimate of Y based on a set of observations y_i, (i = 1, 2, 3, . . .) will be denoted by $\hat{y}$. It is assumed that a specific physical quantity has a unique value if the quantity is sufficiently well defined–if the definition is such that the values of Y that are consistent with it have a range significantly smaller than the anticipated precision of the measurements.

2.2 True value

The true value of a quantity is defined in VIM as a "value consistent with the definition of a given particular quantity" and is described (somewhat circularly) as being "a value that would be obtained by a perfect measurement." Although **a true value** generally cannot be measured, it is still useful to define the concept of a true value of a measurement. When a measurement has been made, the statement of the result in the form

"The value of < *a specification of the measurand* > is $\hat{y}$" ***(I)***

defines the measurand and implies the existence of an unknown value Y, but the stated quantity $\hat{y}$ is only a 'best estimate' of the **true value** of the physical quantity that corresponds to the given specification of the measurand.

2.3 Measurement process

The measurement of a physical quantity is defined by a *measurement procedure.*

- **measurement procedure**: set of operations, described specifically, used in the performance of particular measurements according to a given method [VIM 2.5]

For the purposes of the present discussion, this definition is insufficient, representing only the 'recipe', a 'measurement' being a specific realization of that recipe. What is missing is a definition for the general concept of the realization of the measurement procedure that yields a measurement result. We therefore define:

- **measurement process**: execution of a specified measurement procedure

The **procedure** is the set of instructions that defines the process. The process is the execution of that procedure, but necessarily only an approximation to it. A measurement is a specific example of the measurement process. The inability of the measurement to reproduce the measurement procedure is a source of variability of the results, and a source of error in the measurement, for, under the assumption of causality, observed differences in the results of a measurement process are due to the existence of uncontrolled and uncontrollable variations in the various influence quantities that affect the measurement. If **all** the conditions of the measurement were held at their specified or implied values and **all** the influence quantities were fixed, the results of all repetitions would be the same. [We obviously exclude from this inherently stochastic quantities such as the decay constant of a radio-isotope.]

The distinction between the results of the measurement process and the result of a measurement (a specific execution of the process) is essential for an adequate discussion of the differences between the concepts of 'error' and 'uncertainty.' It is possible and appropriate to analyze the errors of a measurement process, but one never knows the error of an individual measurement. The extent to which the 'process' is not a true realization of the 'procedure' and does not measure the intended measurand generates the error of the measurement, while a lack of knowledge of the differences between the 'procedure' and the 'process' (an inability to know those differences and apply them as a correction to the observed result) contributes to the uncertainty of the value of the measurand. The 'error of method,' the possibility that the measurement procedure, even if followed exactly and with all known corrections applied, does not measure the intended measurand, produces an uncertainty that must be evaluated in non-statistical ways.

Statement *(I)* above is a statement about the value of a specific measurand, and it should be distinguished from related statements that can be made about the measurement process, such as:

"The result of this measurement is $\hat{y}^*$" *(II)*

Statement (I) is a statement about the **measurand**; statement *(II) is* a statement about the **measurement**, and $\hat{y}^*$ may need to be corrected for various systematic effects before a value can be ascribed to the intended measurand.

2.4 Error

The error in a measurement is simply the difference between the measured value and the true value, and since the true value is unknown, so is the error. A measurement of the quantity Y yields a result y. Repeated measurements of the same quantity will in general give different answers and there will be an observed distribution of results y_i, and hence a distribution of errors $\varepsilon_i = y_i - Y$.

The error of measurement is unknown and can only be approximately determined by carrying out additional measurements. But making additional measurements changes and extends the measurement process, and the additional data should then be incorporated into an improved measurement of the quantity Y; the improved value y still has some unknown—although presumably smaller—error.

For the purposes of Error Analysis an error can be expressed as the sum of a *random* error$\varepsilon_{i,r}$ with zero mean and a non-zero variance, and a fixed *systematic* error $\varepsilon_{i,s}$ with zero variance and non-zero mean. In order to specify this decomposition it is necessary to consider 'faithful realizations' of a defined measurement process, requiring that the measurement be repeated under a fixed set of conditions each time in order that the statistical probability distribution remains fixed. By making known modifications to the conditions of the measurement it is possible to explore the distribution of some of the 'locally systematic' errors, but such a statistical analysis is not possible for 'strictly systematic' errors that are fixed and unchangeable.

If the measurement procedure contains an 'error of method,' even an exact realization of the procedure may not realize the definition of the measurand; there may be a fixed 'systematic error' in the measurements that might not be identified by any analysis of repetitive observations. An error or an omission in the specification of the measurement procedure however can also produce variability in the results. A simple example is the determination of the mass of an object without including in its specification a correction for the buoyancy of air. This would produce a error of the order of a few parts in 10^4, but also an uncontrolled variability in the results of the order of several parts in 10^7 arising from variations in barometric pressure and air density.

2.5 Uncertainty

The term 'uncertainty of measurement' has a specific meaning in metrology: uncertainty is the incompleteness of the knowledge of the value of the measurand associated with the given result of a measurement. In the evaluation of a measurement one is presented with a set of observations from which, based on the available information, an inference is to be made about the 'best' value to be assigned to the measurand. The stated result is incomplete unless it is accompanied by an indication of the uncertainty to be attached to the stated value. This uncertainty describes neither the error in the measurement nor the distribution of possible errors, since a given measurement result has only a single, unknown error.

Statement ***(II)*** above is cast in a form that allows one to consider repetitions of the measurement and to ask questions about the statistical distribution of the results; *i.e.*, to consider the errors of the measurement process. Statement ***(I)***, on the other hand, is a statement about a specific result $\hat{y}$; what is not known is the value of the measurand Y. One cannot evaluate the error, but one can ask for a measure of the uncertainty associated with this statement. This quantitative measure is also called 'uncertainty':

> **uncertainty**: parameter, associated with the result of a measurement, that characterizes the dispersion of the values that could reasonably be attributed to the measurand [VIM 3.9]

This definition is a generalization of the definition that appeared in the first edition of the VIM: "an estimate characterizing the range of values within which the true value of the measurand lies." That older definition is equivalent to the quantity that is now called the "expanded uncertainty" or "overall uncertainty," a quantity defining an interval containing the result of a measurement "that may be expected to encompass a large fraction of the distribution of the values that could reasonably be attributed to the measurand," and which is usually taken to be 2-3 times the standard deviation measure of uncertainty.

It is unfortunate that there exists no adequate vocabulary capable of distinguishing between the general concept of uncertainty (doubt, incertitude, the absence of complete knowledge) and the specific parameters (variance, standard deviation, expanded uncertainty) that measure and describe it. The ISO *Guide* includes the statement "Because of the lack of different words for this *general concept* of uncertainty and the specific quantities that provide *quantitative measures* of the concept ... it is necessary to use the word 'uncertainty' in these two different senses."

3 MEASUREMENT UNCERTAINTY

The *analysis* of *error* of a measurement process is based on a statistical distribution, considered as an ensemble of results that could be obtained in repetitions of the measurement process. The distribution of errors is then described in terms of the means and variances of a given process, and, in itself, is unable to describe fixed systematic *errors* that are unchanged in every repetition. These must be treated separately.

The analysis of uncertainty, on the other hand, recognizes that the measurement result is a value $\hat{y}$ that is only an approximation to the value of the measurand Y. 'Uncertainty Analysis' is not primarily concerned with characterizing the statistical distribution of errors, but with characterizing (the limitations of) the information contained in the available data. It attempts to characterize the distribution of values of Y that are consistent with the given estimate $\hat{y}$. The quantitative measure of the uncertainty is the **variance** of that distribution, but it is usually expressed in terms of its square root, the **standard deviation**. Because variance components are additive in the evaluation of the uncertainty of a measurand, we shall here generally use 'uncertainty' to mean a variance rather than a standard deviation. This avoids the awkward and unnecessary 'square root of the sum of the squares' (of quantities that were obtained by taking a square root in the first place). However in discussing specific values it is much more convenient to use the standard deviation, since that has the same dimension as the measurand (and thus may be readily expressed as a fractional value) and a magnitude that can be more easily comprehended than a variance.

The mathematical formulation of the problem can be, and often has been, misinterpreted. The analysis of error describes the distributions of y_i and of $\hat{y}$, given the true values of Y and of whatever 'influence quantities' z_k that affect the observations:

$$\Pr(\hat{y}_i) = f(\hat{y}_i | Y; z_1, z_2, \ldots)$$

where the variables following | mean that Y and z_i, representing all other parameters and influence quantities that affect the determination of the estimate $\hat{y}$, are held constant.

The analysis of uncertainty is concerned with information theory rather than with statistics *per se*. Measurement is a process that increases the information available on the value of Y. The purpose of uncertainty analysis is to describe the information about the value of the measurand that has been achieved as a result of the measurement process, and the distribution of values of Y that are consistent with the available information:

$$\Pr(Y) = g(Y|\hat{y};\ z_1,\ z_2,\ldots)$$

3.1 Bayes' theorem

Bayes' Theorem[3] expresses a relationship between conditional probabilities associated with a multi-component probability distribution. In its simplest form, the joint probability density of X and Y, $p(x,y)$ (for example, height and weight of individuals in a population) can be written in two equivalent ways:

$$p(x,\ y)\ dx\ dy = p_1(x|y)\ dx\ p_2(y)\ dy = p_2(y|x)\ dy\ p_1(x)\ dx$$

where $p_1(x|y)dx$ is the conditional probability of the occurrence of x when it is known that $Y = y$, and $p_2(y|x)\ dy$ *is* the conditional probability of the occurrence of y when it is known that $X = x$.

The unconditioned probability densities of X and Y are

$$p_1(x) = \int p_1(x|y)p_2(y)\ dy \qquad p_2(y) = \int p_2(y|x)\ p_1\ (x)\ dx$$

The function $p_2(y)$ *is* the distribution of y independent of the coexisting value of x. We then obtain

$$p_2\ (y|x) = \frac{p_1(x|y)p_2(y)}{\int p_1(x|y')p_2(y')\ dy'} \qquad \text{Bayes' Theorem}$$

which relates the conditional probability density of y when the value of X is specified to the conditional probability density of x when the value of Y is specified.

From the frequentist point of view there is a defined and observable joint distribution of x and y, and $p_1(x)$ and $p_2(y)$ are the marginal distributions—for each value of x the joint distribution is integrated over all values of y, and for each value of y the joint distribution is integrated over x. There can be no objection to Bayes' theorem when it is stated in this 'pure' form; i.e. when probability distributions are frequentist statements concerning jointly observed variables.

It is when Bayes' Theorem is applied in a 'causal' situation that it has engendered opposition from frequentists. From the information viewpoint, $p_2(y)$ represents what is known about the possible values of y in the absence of any information as to the value of x; it is the 'prior distribution' of one's knowledge of y, before there is information on x. By extension, Bayes' theorem can be invoked to provide the basis for determining estimates of the values of the unknown parameters from a set of observations[4]. The process of measurement, observing values of x_k, changes what is known about the value of Y:

$$g(Y|\hat{y}(x_1,x_2,...)) = \frac{p(Y)f(\hat{y}|Y)}{\int f(\hat{y}|y')p(y')dy'} \qquad (1.1)$$

If we initially know 'nothing' about the value of Y, we set $p(Y) = K$ = constant for all values of Y for which $f(y|Y)$ *is* significantly different from 0. Then

$$g(Y|\hat{y}(z_1,z_2,...)) \approx \frac{f(\hat{y}|Y)}{\int f(\hat{y}|y)dy} \qquad (1.2)$$

and $g(Y|\hat{y})$ is approximately proportional to $f(\hat{y}|Y)$.

3.2 Evaluation of uncertainty

A measurement of a quantity X may yield a set of observations x_i; in its simplest form, the error model for the measurement process is given by

$$x_i = X + \varepsilon_i \qquad (2\text{-}1)$$

where the error ε_i is considered to be characterized by a distribution with mean δ and a variance σ^2. The mean of the observations, $\bar{x} = (1/n)\sum_i x_i$, will approximate the distribution mean, X+δ. Within the measurement process itself, however, there are no observations that can be made to evaluate the 'systematic error' δ. The best estimate of X is given by

$$X \approx \hat{x} = \bar{x} - \hat{\delta} \qquad (2.2)$$

where $\hat{\delta}$ is an estimate of δ that must be determined from information other than that contained in the observation set x_i.

The 'uncertainty of the measurement' is a statement about the uncertainty associated with the identification $X \approx \bar{x} - \hat{\delta}$. That uncertainty can be decomposed into one component associated with identifying $X = \bar{x} - \delta$. and another component associated with identifying $\delta = \hat{\delta}$. The uncertainty assigned to X+δ is based on the analysis of the set x_i, the value of $\hat{\delta}$ and its uncertainty must come from sources of information external to the measurement process, but could be based on the statistical analysis of other measurements.

If the estimate $\hat{\delta}$ is independent of $\bar{x}$, the uncertainty of X is given by

$$u^2(X) = u_{\bar{x}}^2 + u_{\hat{\delta}}^2 \tag{2.3}$$

By Bayes' Theorem, the uncertainty component $u_{\bar{x}}^2$ is $s^2(\bar{x})$ where

$$s^2(\bar{x}) = \frac{1}{n(n-1)}\sum_{i=1}^{n}(x_i - \bar{x})^2 \approx \sigma^2(\bar{x})$$

and $u^2(\hat{\delta})$ is the variance of a subjective distribution of the degree of belief of the quantity $\hat{\delta} - \delta$. This latter quantity is the variance associated with what is not known about the 'systematic error" δ. Both the distribution of $\bar{x}$ and the distribution of $\hat{\delta}$ are subjective distributions in the sense that neither of these distributions is directly observed. The uncertainty components derived from these distributions are therefore to be treated similarly: each contributes a variance component to the total uncertainty.

As implied by the above evaluation, uncertainty can be analyzed in terms of its components. Uncertainty components are equivalent to statistical variances, and more generally to the elements of a covariance matrix u_{jk} ($u_{jj} = u_j^2$,) and combine as variances. When a quantity Y is evaluated from its functional dependence on other quantities ξ_j,

$$Y = F(\xi_1, \xi_2, \xi_3, \ldots),$$

where ξ represents the full set of all such quantities (measured or unmeasured), the uncertainty of Y is evaluated from the uncertainties of ξ_j in terms of the sensitivity coefficients by

$$u^2(Y) = \sum_{j,k}\frac{\partial F}{\partial \xi_j}u_{jk}\frac{\partial F}{\partial \xi_K} = \sum_j\left[\frac{\partial F}{\partial \xi_j}\right]^2 u_j^2 + 2\sum_{k<j}\frac{\partial F}{\partial \xi_j}u_{jk}\frac{\partial F}{\partial \xi_k} \tag{2.4}$$

The uncertainty variances u_j^2 and the uncertainty covariances u_{jk} may be calculated by similar expressions if they, in turn, are derived from other quantities.

The set $\xi = \xi_i, (i = 1,2,\ldots)$ contains all quantities that can be identified as contributing to information concerning the value of Y. The space of this set may be described as being projected into the sum of two spaces, $S=A+B$, where A is the space of those variables whose uncertainties can be determined by statistical means [Bayes' Theorem, relating $\Pr(\hat{\xi}_i|\xi_i)$ to $\Pr(\xi_i|\hat{\xi}_i)$], and B is the space of those variables whose uncertainties must be determined by other means, but both classes of uncertainty evaluations are to be expressed in terms of the dispersion (second central moment) of the putative distribution of a degree of belief.

All uncertainties are treated equally, whether ultimately evaluated from information based on a statistical evaluation of repeated observations or not. The uncertainty variance $u^2(Y)$ is the sum of the variance components of all the terms that contribute to the lack of information as to the 'true' value of Y. Uncertainties are, in the strict sense, unobservable since they represent the dispersion of a degree of belief in the unknown 'true' value of a physical quantity, and not directly the dispersion of data. Those components that can be evaluated from the statistical analysis of observed results are, of course, determined on the basis of that analysis, using Bayes' theorem to transfer the variability of the observed data to the uncertainty in the value of the physical quantity.

KEYWORDS

Metrology, measurement, uncertainty, error, covariance, ISO guide

REFERENCES

[1] International Vocabulary of Basic and General Terms in Metrology, (2nd edition), International Organization for Standardization [published in the name of BIPM, IEC, IFCC, ISO, IUPAC, IUPAP, OIML] (1993).

[2] Guide to the Expression of Uncertainty in Measurement, International Organization for Standardization [published in the name of BIPM, IEC, IFCC, ISO, IUPAC, IUPAP, OIML] (1993).

[3] Bayes T. , An Essay towards Solving a Problem in the Doctrine of Chances, Phil.Trans. Roy. Soc. 53,370 (1763); G. A. Barnard, Biometrica, 45, 293 (1958).

[4] Weise K., Woger W., A Bayesian Theory of Measurement Uncertainty, Measurement Science and Technology, 4, 1 (1993).

FUZZY AUTOMATA AND GRAMMARS : MANIPULATION OF UNCERTAIN KNOWLEDGE

Claude GAUDEAU[1], Vero RAKOTOMALALA[2], Laurent GOUTHIERE [2,3] Eric RAVAUD[3] and Stéphanie BENOIST[2]

[1] *CNRS - Laboratoire de Bio-Informatique, 32 rue Emile Zola, 37000 Tours, France.*
[2] *Société de Bio-Informatique et de Bio-technologie, 45 rue Emile Zola, 37000 Tours, France.*
[3] *CNAM, 2 rue Léo Delibes, 37200 Tours, France.*

ABSTRACT

The concepts of fuzzy automata and grammars allow the modelization of new potentialities in interconnected biomolecular systems. A Fuzzy Automata is defined by the quintuplet A = (U, Y, X, f, g,) whereas fuzzy grammar can be defined by an alphabet, an axiom, and a group of fuzzy rules, and can be expressed through linear equations. These concepts can be applied, for example, to the indexation of scientific documents. A hypothesis of associative auto-organization has been established suggesting that emergence represents the actualization of potentialities. After its decomposition into fragments, the mathematical modelization of the emergence of new properties is made possible. This can be demonstrated subsequently through fuzzy systems. The procedures of modelization and simulation are closely associated in an inference system : the expert system.

RESUME

Les concepts des automates et des grammaires flous permettent la modélisation des nouvelles potentialités dans des systèmes biomoléculaires interconnectés. Un automate flou est défini par le quintuplé A = (U, Y, X, f, g,), tandis que les grammaires floues peuvent être définies par des équations linéaires. Ces concepts peuvent être appliqués, par exemple, à l'indexation automatique des documents scientifiques. Nous avons établi une hypothèse de l'auto-organisation associative. Celle ci suggère que l'émergence représente l'actualisation des potentialités. Après avoir été décomposée en fragments, la modélisation mathématique de l'émergence des nouvelles propriétés est rendue possible. Ces propriétés peuvent être par la suite démontrées à l'aide de systèmes flous. Les procédures de la modélisation et de la simulation sont toutes deux étroitement liées dans un système d'inférence: le système expert.

1 INTRODUCTION

In order to take into account the imperfections in scientific knowledge which can be obtained from experimental data, it is necessary to introduce concepts developed in fuzzy logic into models, more precisely the concept of fuzzy automata and grammars. These concepts lead to the definition of a fuzzy system (Kaufman, 1973, Kabbara, 1979).

This approach can allow one to modelize the potentiality and emergence of new properties of interaction in quantum mechanics or quantic electrodynamics and interconnected bio-molecular systems (Redon, 1983). It allows the representation of biological functions and the description of the complexity of self-organized systems (Chauvet, 1990). Generally, the method used in systems takes into account only those parts of the system which can be controlled at the input level and those that can be observed at the output level. An expert system of modelization allows a complex system to be described by means of the SADT (structural analysis and design techniques) method. It can answer questions relating to imprecise information concerning each elementary system borrowed from expert and scientific publications stored in the system's knowledge base (Gaudeau et al, 1983).

Much of this knowledge is vague. The inference procedures are enhanced with mechanisms that combine degrees of evidence according to the rules of plausible reasoning, which in turn draws conclusions from facts that appear correct. In this case we use approximate reasoning, which allows conclusions to be drawn by taking into account the degree of consistency in the facts. One use of this method is to apply it to modelization and simulation of interaction between macro-molecular systems and physical and chemical agents. However, in principally biological systems, this method is not complete because of existing potentialities or latent properties in isolated or elementary systems. These properties, which cannot be at first observed, may appear when several systems are associated. In this case, there is the emergence of new potentiality (Quasther, 1970). In order to modelize and simulate complex systems, it is necessary to use expert systems (Gaudeau, 1993).

2 BASIC CONCEPTS

2.1 Fuzzy automata

Definition of Fuzzy Automata :

A Fuzzy Automaton is defined by the quintuplet :

A = (U, Y, X, f, g) where:

- U is an unempty finished set of input elements
- Y is an unempty finished set of output elements

- X is an unempty finished set called internal states
- f is the belonging function of a fuzzy subset of X x U x Y ,meaning that an application of X x U x Y in the closed set [0,1] ∈ R, f is called the fuzzy function of direct transition (f.f.t.d).

- g is the belonging function of a fuzzy subset of Y x U x X ; thus an application of Y x U x X in the closed [0,1] ∈ R , g is called the fuzzy function of direct output (f.f.d.o.) (B. Conche, 1974) :

Remarks : U, Y, X *are finished sets, in the classic definition of the term;* f *and* g *are property functions of finished fuzzy subsets.*

Operation of fuzzy automata

Diagram of the automata operation :

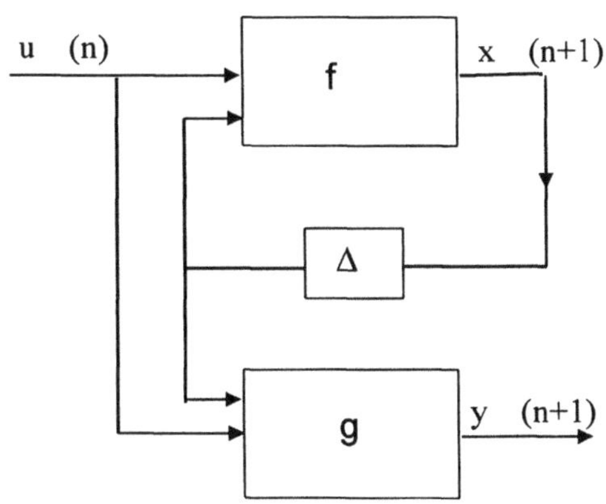

Figure 1

n, (n+1) represent discreet elements of time,
u (n) ∈ U is the input element of the automata at the moment n
x (n+1) ∈ X is the internal state of the automata at the moment (n+1)
y (n+1) ∈ Y is the output element at the moment (n+1)
Δ is the delay function (passage from moment n to moment (n+1)

2.2 Fuzzy grammar

2.2.1 Definition

A fuzzy grammar is defined by :

1 - An alphabet for example :

input	→	$U = \{ u_1, u_2\}$
state	→	$X = \{x_1, x_2, x_3, x_4\}$
output	→	$Y = \{y_1, y_2 \}$

2 - <u>An axiom</u>

A (0) C X (fuzzy state subset)

3 - <u>A group of fuzzy rules</u>

defining a fuzzy relation R defined upon X x U x X, of which the membership function is

$$(\mu_R (x_i + 1) \mid x_i , u_j) \in [0,1] \; ; x_i \in X, u_j \in U$$

The rules can be associated with value V such as $V \in [0,1]$

2.2.2 Example

For the alphabet { X, U} defined $x = \{ x_1 , x_2 , x_3 , x_4 \}$

$u = \{ u_1, u_2 \}$

	x_1	x_2	x_3	x_4
Axiom : A (0)	0,3	1	0,4	0

Rules of grammar :

$$\text{set} \begin{bmatrix} x_1 \\ \\ x_2 \\ \\ x_3 \\ \\ x_4 \end{bmatrix} = \begin{bmatrix} 0,3 & 0 & 0,4 & 1 \\ 0 & 0,2 & 0,8 & 0 \\ 0,3 & 0,5 & 1 & 1 \\ 0,8 & 0,9 & 0 & 0,4 \\ 0,5 & 0,9 & 0,1 & 0,3 \\ 0,3 & 0,5 & 0,5 & 1 \\ 0,3 & 0,7 & 0 & 1 \\ 1 & 0,2 & 0,3 & 0 \end{bmatrix} \begin{bmatrix} x_1 \\ \\ x_2 \\ \\ x_3 \\ \\ x_4 \end{bmatrix} {*}{=}\mu_R (x_3 \mid x_4 \begin{bmatrix} u_1 \\ u_2 \end{bmatrix})$$

Remark: the + operator of the matrix correspond to OR ($\cup$)

The fuzzy grammar explains the function of the fuzzy automata in the form of rules which, in certain conditions, can be expressed by a linear system of equations.

For the input sequence (u_2,u_1, u_1,u_2) we obtain, for instance :

step	input				
1	u_2	0,8	0,9	0,4	0,4
2	u_1	0,4	0,5	0,9	0,9
3	u_1	0,5	0,9	0,5	0,9
4	u_2	0,9	0,9	0,5	0,5

at step 4, the automata is in a composite state where states x_1 and x_2 have a membership coefficient
$\mu_{A4}(x_1) = \mu_{A4}(x_2) = 0{,}9$

At this state, the matrix can be associated with two graphs corresponding to :

- the command U^1
- the command U^2

which are represented in Figure 2.

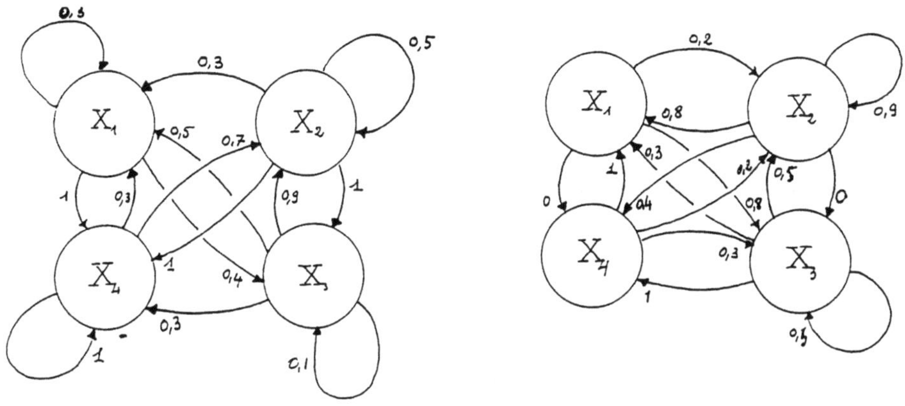

U^1 U^2

Figure 2: Graph Representing the grammar of the fuzzy automata for commands U^1 and U^2.

2.2.3 Applications

The concept of fuzzy automata and grammars can be used for automatic indexation of scientific documents :

1. To each of the key words can be assigned a fuzzy value V= [0,1], which weighs the pertinence of the key word in the text. This can be obtained by automatically evaluating the importance or the occurrence of the word in the document.

2. To each couple of key words can be assigned a value representing the importance (or occurrence) of the relation between two key words. A standardization of scientific documents could be eventually proposed through INTERNET.

Retrieval Procedure :

The document will then be indexed by a n x n matrix with appropriate labels for the models of the associated graph.

A documentary system using this method can be improved in order to avoid a surplus or a lack of references. Rules of fuzzy logic can thus introduce a new instrument into the retrieval procedure.

Remarks : The behavior of the fuzzy automata and its associated grammar is analyzed rather as a global action then a sequence of inputs.

Why use the notion of fuzzy automata for automatic indexation using a graph ? As for human intelligence , the meaning of a word or relation is not completely retained at the moment of its appearance in the text, but its comprehension gradually becomes more precise by the conjunction of other words and the global interpretation of the sentence.

2.3 Fuzzy system

$$\mathbf{S = < x, u , f, A_0 , g >}$$

Let :

$X = \{ x_1....x_n\}$ finite states (invariable time)
$U = \{u_1...x_m\}$ finite states of inputs or commands (invariable time)
$Y = \{y_1...y_p\}$ finite outputs

$A_o \in X$ initial state;

$f: X \times U \longrightarrow X$, fuzzy state function
$g: X \times U \longrightarrow Y$, fuzzy output function

N: final time
at $t \in [0, n-1]$ S is submitted to the input $u(t) \in U$.

If Non-Deterministic system :

If the system is non-deterministic, it will be described by :

$A(t+1) = F(x(t), u(t))$ next state function
$B(t) = G(y(t), u(t))$ output function

with

$A(t+1) \in P(x)$
$B(t) \in P(y)$

where P (X) and P (Y) are the sets of the subset of X and Y.

2.3.1 For a fuzzy system the following state equation is considered : $\mathbf{f : X \times U \longrightarrow X}$

Given state x(t+1) at the moment (t+1), knowing that it was in state x(t) at the moment (t) and submitted at input u(t), this means that A(t+1) in the non-deterministic system becomes $\underset{\sim}{A}$ (t+1) in the fuzzy set of X.

F is replaced by the fuzzy relation $\underset{\sim}{R}$. The belonging function associated with $\underset{\sim}{R}$ is :

$(\mu_R(x(t+1)) \mid x(t), u(t), \in [0,1]$

The relation $\underset{\sim}{R}$ represents a fuzzy transition relation and the equation gives the degree of the succession of state x (t + 1) , knowing that the system was at state x(t) and submitted to u(t) at time (t).

At each moment, the system, rather than being classified in a well-determined state, is in a fuzzy subset of states $\underset{\sim}{A}(t)$ with degrees of membership varying from 0 to 1.

The following membership function has to be determined (Kabarra, 1979) :

$$\mu_{\underset{\sim}{A}(t+1)}(x(t+1) = \bigvee_{(x(t),u)\in XxU} \left[\mu_{\underset{\sim}{A}(t)x\{u(t)\}}(x(t),u) \wedge \mu_{\underset{\sim}{R}}(x(t+1)|x(t),u)\right]$$

and more generally

$$\mu_{\underset{\sim}{A}(t+1)}(x(t+1) = \bigvee_{(x(t)\in X} \bigvee_{u\in U} \left[\mu_{\underset{\sim}{A}(t)}(x(t)) \wedge \mu_{\{u(t)\}}(u) \wedge \mu_{\underset{\sim}{R}}(x(t+1)|x(t),u)\right]$$

where u(t) is the membership of a fuzzy subset at each time.

2.3.2 Output function : g : X x U ⟶ Y

Idem, the output function of the non deterministic system B(t) becomes B and G is represented by the fuzzy relation T with the belonging function μ_T : from which -

$$\mu_{\underset{\sim}{B}}(y(t) = \bigvee_{x(t)} \left[\mu_{\underset{\sim}{A}}(x(t)) \wedge \mu_T(y(t))|x(t),u(t))\right]$$

3 Potentiality and Emergence Modeling in Interconnected Systems

Modelization consists of isolating an abstract, reduced version , permitting the simulation of the real universe. Confrontation with experimental results allows validation of the model.

3.1 Notion of potentiality and emergence

Particularly in biological systems, Chauvet (1990) elaborated a hypothesis of associative auto-organization which implies the notion of potentiality and emergence of systems which can be expressed in the following terms :

«If at a given moment a structural elementary unit (or system) Ui no longer assures a physiological, elementary, necessary function fi, it must then, in order to survive, receive this function from another structural unit which possesses it.»

This implies that there exists at the potential state a new functional interaction. A second definition was introduced by Chauvet (1990) :

« A level of functional organization is constructed upon the subset of structural units (or systems) which have the same potentials and of which the collective behavior defines this function.»

Emergence represents the actualization of potentialities.

3.2 Notion of system fragments

The concept of potentiality (non-observable) necessitates the breakdown of an elementary system into fragments (Gaudeau et al, 1988; Crendal, et al, 1970).

3.2.1 System fragments (Figure 3)

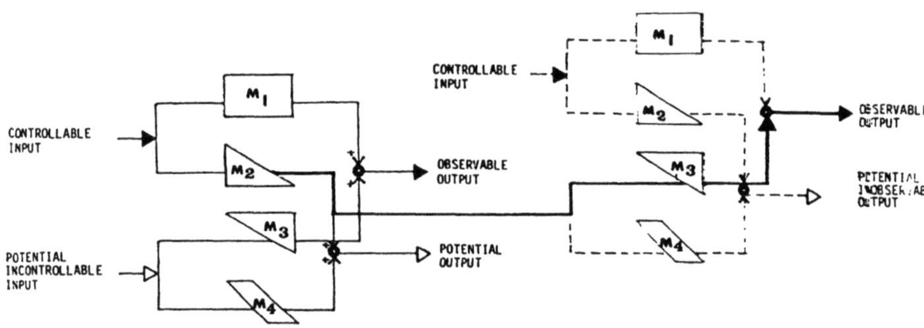

Figure 3: Fuzzy set theory can simulate the association of two molecular systems, broken down in fragment and connected in series, in order to estimate the potentialities of each one (I. Redon: Cell, Pergamon Press, p.32, 1983).

A system S can be decomposed into four fragments M_1, M_2, M_3, M_4. (Gaudeau, et al, 1979, 1988).

M_1 can be observed and is controllable.
M_2 cannot be observed but is controllable
M_3 cannot be controllable but is observed
M_4 is neither observable nor controllable.

3.2.2 Association of system fragments

The association of several systems will show clearly that some observable variables depend on non-observable variables.

Consider two compatible systems, S and S'. Let us associate the non-observable part of S to the non-controllable part of S' as shown in Figure 3.

The bold-lined path shows that the association of two fragments M2 of S and M3 of S' allows the demonstration of a phenomenon which at first was only potential, i.e. the inobservable output of system S.

<u>Example :</u>

The following system and its vectoral equation are decomposed into four parts : M1, M2, M3, M4, as defined earlier :

$$\begin{bmatrix} M_1\ (t+1) \\ M_2\ (t+1) \\ M_3\ (t+1) \\ M_4\ (t+1) \end{bmatrix} = \begin{bmatrix} -1 & 0 & 1 & 0 \\ 2 & 1 & 0 & 1 \\ 0 & 0 & 1 & 0 \\ 0 & 0 & 0 & -2 \end{bmatrix} \begin{bmatrix} M_1 \\ M_2 \\ M_3 \\ M_4 \end{bmatrix} + \begin{bmatrix} 1 & 2 \\ 1 & 1 \\ \boxed{0 \quad 0} \\ \boxed{0 \quad 0} \end{bmatrix} \begin{bmatrix} e_1 \\ e_2 \end{bmatrix}$$

→non-controllable

$$S = [\,1\ 0\ 1\ 0\,] \begin{bmatrix} M_1 \\ M_2 \\ M_3 \\ M_4 \end{bmatrix}$$

non observable

The breakdown into fragments permits one to mathematically model the possible emergence of new properties. This method can be applied to several domains, particularly the biomedical, through the compartments theory and, more specifically, quantic electrodynamic and molecular biophysics.

3.3 Association of fuzzy systems

We then assume that there exists in the fuzzy system, at each time, a dependence between the state and the inputs : $\underset{\sim}{Q}(t)$.

We associate two fuzzy sets $\underset{\sim}{W}$ and $\underset{\sim}{W}$ to reach a shared goal. We assume that $\underset{\sim}{W}$ and $\underset{\sim}{W}$ have fuzzy inputs parameterized by states. This is represented by the fuzzy relations $\underset{\sim}{Q}(t)$ and $\underset{\sim}{Q}'(t)$ respectively for $\underset{\sim}{W}$ and $\underset{\sim}{W}$. We also assume that the state of W influences W', and vice versa. We represent this influence by the fuzzy transition relation $\underset{\sim}{R}(x(t))$ for $\underset{\sim}{W}$ and $\underset{\sim}{R}'(x(t))$ for $\underset{\sim}{W}$. The association of $\underset{\sim}{W}$ and $\underset{\sim}{W}$ is represented as follows :

$$\mu_{\underset{\sim}{A}(t+1)}(x(t+1)) = \underset{u(t)}{\vee}\ \underset{x(t)}{\vee} \left[\mu_{\underset{\sim}{Q}(t)}(x(t),u(t)) \wedge \mu_{\underset{\sim}{R}}(x(t+1)|x(t),u)) \right]$$

We have the same equation with A', W', Q', and R'.

The two systems will try to evolve on shared states to reach their goal. Thus, they must be compatible. In the best case, the fuzzy sets of W and W' are the same. Compatibility can be expressed by Hamming's distance, which is the distance currently used in the fuzzy sets theory.

The concept of potentiality is then introduced at three levels :

Level 1 : the states. X is the set of all the possible states of W and W' as well as those that may be reached during the experiment.

Level 2 : Inputs. There is a dependence between the inputs and the states.

Level 3 :W and W' influence W', and reciprocally. It is interesting to introduce the fuzzy systems to demonstrate potentiality, but the implementation is difficult and complex.

4 Expert System of Modelisation and Simulation

After having defined modelization, the object of simulation is to analyze the consequences of a model or a theory, to infer the new facts, to predict qualitative behavior (generally formalized in the form of « state »), and to give quantative responses (Gaudeau et al, 1993).

Thus, the procedures of modelization and simulation are closely associated in an inference system : the expert system.

4.1 Modelization and simulation expert system SEI4

A basic expert system consists of :

1. a language of knowledge expression
2. a fact base
3. an inference model (0 + order).
4. a knowledge base

which form the ensemble from which the system makes its decisions, which then regroups the input and output and the control values of manipulated elementary systems.

The system is broken down into functional blocks by the SADT method (Fig. 4). To each system is associated a rule in which the premises correspond to the input and to the control variables and the conclusion to the output. To this rule is associated a function which permits the manipulation of quantitative data associated with the inputs, controls, and outputs.(Gaudeau et al 1993)

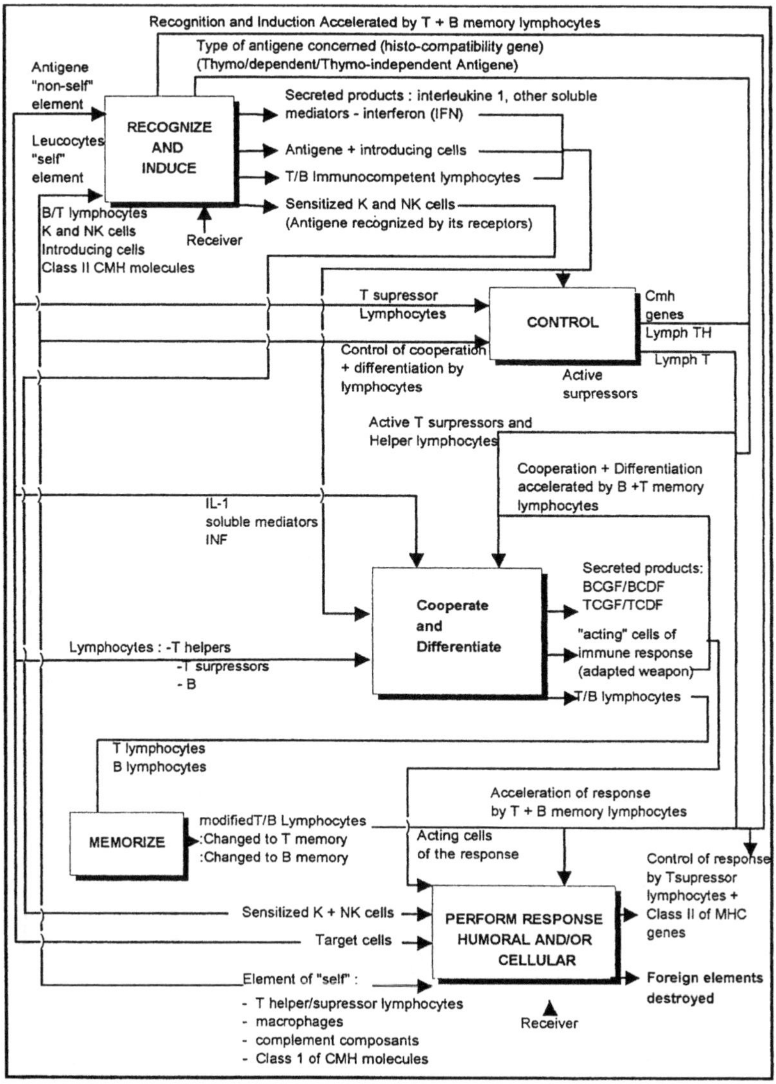

Figure 4. Representing the first SADT level of the immune response model by an ensemble of actigrams or functional blocks. Each block (RECOGNIZE AND INDUCE, COOPERATE AND DIFFERENTIATE, PERFORM, RESPONSE) is decomposed into other functional blocks. Each, with its inputs and outputs, is translated into pseudo-natural language in the form of one or several rules according to the number of inputs.

This expert system was applied to the modelization of immunological reactions and of infection by the HIV virus. (Figure 5)

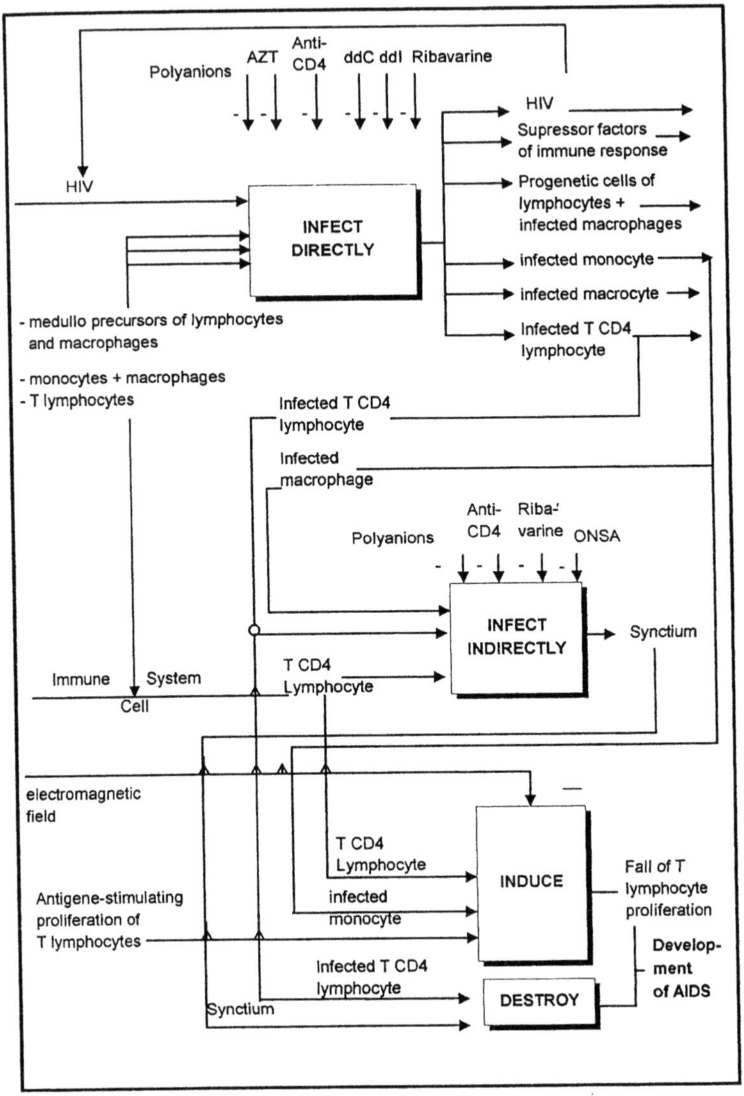

Figure 5 : Representing the first level of SADT of the immune system infected cells by the HIV virus for modelization and simulation by the expert system . Each functional block will generate one or several rules of the system. Therapeutic actions are represented by control variables (AZT, Anti-CD4, EMF).

4.2 Fuzzy expert system

A second expert system which functions in fuzzy logic was elaborated where each fact is associated with the fuzzy value v = [0,1] and each rule is associated with a level of influence I = [0,1]. This method allows the connection of elementary systems among themselves only if the threshold of release is superior to a certain value d. This technique permits the simulation to function at different levels. The next step is to control the functional blocks represented by fuzzy systems. The latter requires the configuration:

- PC compatible with at least one microprocesser INTEL 80286
- MS-DOS version 2.11 or superior, with at least 512 Ko of RAM memory
- Color Screen (VGA, SVGA)
- Mouse or keyboard and directional arrows

and the general characteristics :

- Expert system with O order chaining before integration of fuzzy logic
- Source language C to assure portability to other environments
- Displacement by rolling menus
- Contextuel aid
- Ability to manage large volumes of rules and facts (1,000 rules and 5,000 facts).
- Integrated sample rule bases

CONCLUSION

The fuzzy system is a new method allowing for the representation of uncertain knowledge and improved modelization of systems and their interconnection (Sanchez, 1976).

Theories of potentiality and the emergence of potential systems, which have been developed by several authors, can be applied to physical and molecular systems by using a decomposed algorithm in order to improve knowledge of the properties of the subunits.

For instance, in the assembled molecule, the subunit structures and properties are not the same as in the isolated molecule. If new intermolecular interactions occur, this will allow us to redefine the model.

Automata theory has already been applied to molecular interactions by Atlan. New developments of this theory can also allow one to describe the latent properties of molecular subunits, in particular the concept of fuzzy automata.

Futhermore, the expert system brings a new method of assembling the systems between us and the simulation of new properties. Thus, when dealing with complex systems including numerous reactions, such as the immunological system, we can assume that this method offers a highly effective syntax instrument.

ACKNOWLEDGEMENTS

We would like to thank Ms. Martine Mercier for her contribution to this article.

KEYWORDS

Fuzzy logic, fuzzy automata, fuzzy grammar, fuzzy system, modelization, emergence, expert system.

REFERENCES

[1] Atlan H., Ann. Phys. Biol. Med 1, 15 (1970).
[2] Chauvet G., Traité de Physiologie Théorique - 3, Physiologie intégrative : champ et organisation fonctionnelle, Masson, Paris (1990).
[3] Conche B., Apprentisage Heuristique en Selection de Projets de Recherhe : Une Application des Concepts Flous à la Classification Automatique - Thèse de Doctorat 3ème Cycle, Université de Paris IX Dauphine (1974).
[4] Gaudeau C, Bosma M., Rossazza J.P., Spatial modelling and simulation of organized macromolecular structures by means of factor analysis of electron microscopy pictures, Molecules in Physics, Chemistry, and Biology, I, 129-157 (1988).
[5] Gaudeau C., Crendal W., Vincent S., Galea J.M, Laforge J.M., Notion de fragment et de territoire dans l'identification des Bio-systèmes complexes.R.B.M, Vol 1., N° 5, 377-379 (1979).
[6] Gaudeau C., Rakotomalala V., Asseray R., Charra M., Ferhi F., Système expert de modélisation de la chimiothérapie et de l'électrothérapie dans les maladies induites par le virus de l'immunodéficience aquise, XIVème Congrès International de Biologie Mathématique, Sept.(1993).
[7] Kabbara G., Optimisation floue - Thèse de mathématiques appliquées, Université de Provence, 28 June (1979).
[8] Kaufman A., Introduction à la Théorie des Sous-ensembles Flous - 1. Eléments théoriques de base, Masson, Paris (1973).
[9] Quasther G., Emergence in Biological Organizations, Yale University Press(1970).
[10] Redon I., Morand D., Gaudeau C., Emergence and potentiality in interconnected systems. Fuzzy information, knowledge representation and decision analysis, Congrès IFAC, Pergamon Press, July (1983).
[11] Sanchez E, Roland S., Relations floues, Fonctions floues, Application à l'aide au diagnostic en pathologie thyroïdienne, Medical Data Processing Symposium, Taylor and Francis Ltd (eds) Toulouse (1976).

Data Processing in Chemistry with Neural Networks : An Alternate Way in Cases of Fuzzy Data or Incomplete Models

Jean-Pierre DOUCET, Annick PANAYE and Gilles MATHIEU

Institut de Topologie et de Dynamique des Systèmes, associé au CNRS URA 34, Université Paris VII, 1 rue Guy de la Brosse, 75005 Paris, France

Abstract

Artificial neural networks (ANN) show attractive capabilities for providing non linear mapping schemes and extracting significant features from raw data when the underlying mechanisms are not yet well known. Some applications in basic problems of spectrochemical analyses are discussed, in cases where some fuzziness remains in the data to be processed or the models to be used. Special emphasis is put on property prediction: Examples are given for ^{13}C NMR shift calculation, illustrating the efficiency of the approach.

Resume

Les réseaux de neurones artificiels ont fait la preuve de leur capacité à traiter des représentations non linéaires et à extraire les éléments significatifs d'un ensemble de données brutes, lorsque les mécanismes sous-jacents sont encore mal connus. Quelques applications sont présentées dans le domaine de l'analyse spectrochimique dans des cas où un certain flou demeure quant aux données à analyser ou aux modèles à employer. Un intérêt particulier est accordé à la prévision de propriété. Les exemples présentés dans le calcul des déplacements chimiques en RMN ^{13}C illustrent l'efficacité de cette approche.

1 Introduction

Spectral analysis in Chemistry is often faced with the treatment of noisy data with ill-defined or incomplete models. This intervenes for instance when extracting reliable data from overlapping signals, determining the concentration of an analyte in a multicomponent mixture or predicting spectral parameters (NMR chemical shift, IR vibration frequency...) for an unknown. We examine here what help neural networks can provide in that field in cases where some fuzziness remains in the data to be processed or in the knowledge to be used. Special emphasis will be put on property prediction.

Neural networks in Chemistry are still in a pioneering phase. Software simulations (Artificial Neural Networks "ANN") are still far from the performances to be expected from hardware implementation on dedicated chips. However they yet offer very promising avenues, particularly for their capability of mapping non linear data and extracting complex relationships when underlying mechanisms are still not well known (1,2).

Up to now, most applications rely on layered feed-forward networks. Such networks are composed of elementary units (neurons) organized in layers. Typically, each unit of a given layer is connected to each unit of the preceding and of the following layers with connections of adjustable strengths (connection weights). The input layer is fed with parameters characterizing the problem: in property prediction, such as NMR shift prediction (3,4), they may be structural descriptors (substituent parameters, topological codes, quantum indices..) specifying the environment of the property focus (i.e. site or group responsible for the property sought for), in titration problems, absorbance values at selected wave numbers may be used (5). One (or sometimes several) hidden layer ensures the encoding of the underlying information and allows for non-linear mapping. An output layer delivers a scaled value of the investigated property. As to the elementary unit, the neuron, it receives information from all units of the preceding layer, scaled by the connection strengths. It sums up these inputs and transforms them, thanks to a transfer function (generally a sigmoïd) into an output value which is transmitted to the units of the following layer (a bias allows for shifting the mid point of the transfer function). Figure 1.

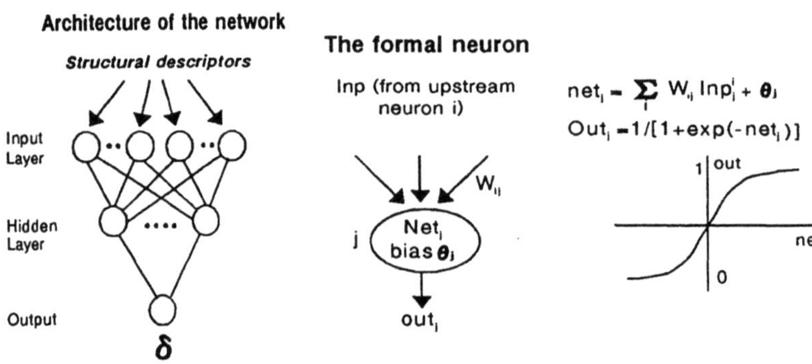

Figure 1. Formal neuron and architecture of a layered feed-forward network. The input layer here comprises 12 neurons with binary inputs (1 if the corresponding position in the environment of the ^{13}C is occupied, 0 otherwise). The output layer (one neuron) delivers a scaled value for the chemical shift. Best results are obtained using seven or nine units (when the "twin descriptors" are used) or six-seven for the "unique" structural description.

using seven or nine units (when the "twin descriptors" are used) or six-seven for the "unique" structural description.

The network operates in a supervised mode. In a first (training phase) the network is fed with a series of pairs (structural descriptors as inputs/observed property values as outputs) constituting the learning set. Connection weights and biases are adjusted so that the network can calculate output values satisfactorily matching the observed values. This adjustment is carried out thanks to the algorithm of "back-propagation of the error". After this training phase, connections and biases are frozen. The network is ready to calculate the investigated property for a new compound whose structural descriptors are given as inputs. The predictive ability is checked on a test set formed with compounds not included in the learning set. This separation of a learning and a test set avoids overtraining the network and allows for maintaining its predictive ability for unknown structures (1,2).

Although this work is focused on the prediction of spectroscopic parameters (in fact ^{13}C NMR shifts), other important applications of ANN in Spectroscopy must be recalled.

2 PEAK PICKING, SPECTRAL CALIBRATION AND QUANTITATION

For instance, it was shown that ANN, using a movable window scanning a spectrum, are able to extract individual peak profiles from overlapping signals (6,7). Results, albeit not as good as from a human expert, yet appear superior to those obtained from the common "second derivative" methods.

ANN were also recently applied to spectral calibration and quantitation, a critical step in spectrochemical analysis (5,8,9). Generally speaking, analysis of multicomponent mixtures is carried out assuming some kind of linear relationship between concentrations and signal intensity (as in the well known Beer's law), and using methods based on linear algebra, such as Multiple Linear Regression, or more recently Partial Least Squares regression, Principal Component Regression...(5).

However, random errors in concentration of standard samples, instrument response or molecular interactions ("matrix effects") result in non linearity. Some methods have been proposed to correct these deviations, such as Quadratic Principal Component Regression or Polynomial Partial Least Squares regression. But they treat non linearity using only simplified polynomial expansions, and appeared only of limited efficiency. The ability of ANN to model non linear, multivariate data makes them very attractive tools in that field. Various recent papers (5,8,9) report the use of layered ANN fed with absorbances at selected frequencies for delivering the concentration of an analyte, after training on calibration standards.

This approach of increased modeling power seems definitely more efficient than linear methods: A factor two is generally gained on the standard error of prediction,

"SEP", but with some cost: the time necessary to train the network. However, it was shown that this drawback can be significantly reduced using some pre-processing: A preliminary Principal Component Analysis, for instance, allows for reducing the number of inputs (and so of connections to adjust in the network). Decreasing the complexity of the network speeds training and makes it easier.

3 Spectral Interpretation

Recognition of structural features from spectra was also tackled via ANN (10,11). Indeed, if IR vibration spectra are quite structure-specific, underlying rules are still largely unknown, particularly in the "fingerprint region" of the spectra. It was established that a neural network fed with an IR spectrum (described as a list of intensity values for selected wave-number ranges) was able to recognize particular functional groups or structural fragments.

Interestingly, examining the functioning of the network, it was shown that the network can by itself build up some new rules from the raw data. For example, in an application devoted to recognition of the substitution pattern of aromatic derivatives (10), the presence of a band in the 857-887 cm^{-1} range was proposed by the network as characteristic of a meta-substituted benzene ring.

4 Property Prediction

Property prediction can sometimes be carried out thanks to theoretical models (such as those derived from quantum chemistry methods). However, up to now, a huge amount of work still relies on empirical models, as evidenced by numerous illuminating examples in ^{13}C NMR shift prediction. ^{13}C shifts are very sensitive to the characteristics of the environment of the resonating carbon, making ^{13}C NMR a very powerful tool in structural analysis. Up to now, due to the lack of a suitable theoretical model of extended structural scope, shift prediction is mainly carried out by means of (empirical) additive increments and frozen substituent constants. However, these models suffer two main drawbacks:

- the need for splitting data into separate subsets, one for each connectivity of the resonating carbon;
- the need for introducing "interaction terms" to take into account the non additive behaviors observed for branched structures or for substituent-induced shifts (3,4 and references therein).

The use of refined topological descriptors, more precisely specifying the environment of the resonating carbon, and evolving substituent parameters gives in the simplest cases improved results (12). But the need for subsets of given connectivity remains. So we investigated the use of neural networks for ^{13}C NMR shift prediction.

The first example presented here concerns methyl substituted cycloalkanes (3). Some among these compounds (cyclohexanes) were previously treated with a (rather intricate) additive model involving numerous interaction terms not easily rationalized (13,14).

In our ANN treatment (3), the environment of the resonating carbon is described specifying the position of the methyl substituent (α,β,γ,δ...) and its orientation (axial or equatorial). In fact, starting from the ^{13}C atom, the cycle may be described in one direction or the reverse one, leading to "twin structural descriptors". However, a unique ordered descriptor (provided with some priority rules) gives nearly equivalent results. Another problem arises for the possibility of conformational interconversion. From the data set of Dalling and Grant (13,14), a first learning set was constituted from carbon atoms not involved in the interconversion process (i.e. 80 δ, gathered in set A of Fig. 2) and a test set from the other ones (7 δ of set B, Fig. 2): For these carbons, shifts are calculated for the two conformers and their mean value compared to the experimental shift. Results are quite satisfactory with mean errors of 0.3 (learning set) and 0.2 (test set) ppm. However, trying prediction on more heavily substituted compounds from a subsequent study (15) (14 δ of subset C) leads to a larger error (0.8 ppm), presumably due to the presence of δ-syn di axial interactions not previously encountered by the network, and this even if carbons directly involved in such interactions are excluded.

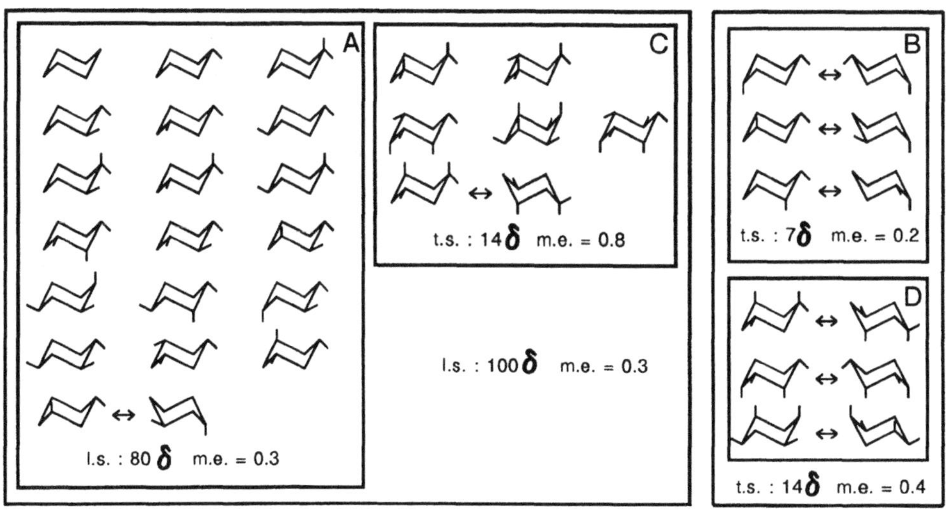

Figure 2. The various data sets used in the treatment of methyl substituted cyclohexanes (see text)

In a second treatment, the learning set was extended by inclusion of these compounds, so that the network can learn such interactions (learning set: A plus C, that is 100 δ values, including carbons involved in δ-syn di axial interactions). Results obtained on a test set gathering interconverting structures of sets B and D, are now quite satisfactory with a mean error of 0.4 ppm for 14 δ values (mean error of the learning set: 0.3 ppm on 100 δ). For the sake of comparison it must be recalled that the additive incremental system of Dalling and Grant, plus a new increment (7.2 ppm) in order to take into account α axial/γ axial interactions, leads to a mean error of only 0.5 ppm for the whole set of data (114 values). (Figure 2).

In an exploratory way, we tried to extend this approach to methylated five- to seven-membered cycles (considering only secondary and tertiary carbons). By analogy with the treatment of cyclohexanes, substituents are described as "axial" or "equatorial", although, obviously, this constitutes only a very rough approximation. The network is similar to that used for cyclohexanes, except that we now have to specify if the cycle is five-, six- or seven- membered, amounting thus to 16 input units. Owing to the increased number of connections it would be difficult to select a learning and a test set. So we use a cross validation process ("leave one out" process): successively one shift value is excluded from the training set and is re-calculated after training the network. Although not so good as for methyated cyclohexanes, results appear quite promising. Some examples are given in Figure 3.

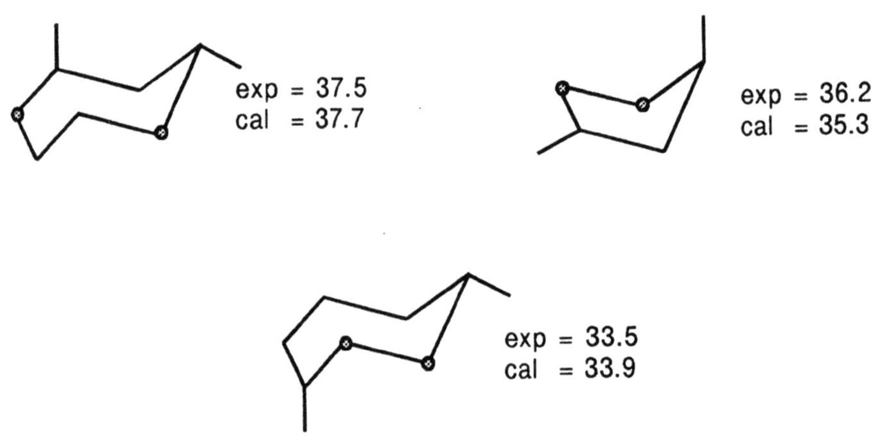

Figure 3. Calculated and Observed ^{13}C shifts for some five- or seven- membered cycles; the ^{13}C are indicated as grey circles.

Another example in ^{13}C NMR relies on the evolvement of substituent shifts (shifts induced by direct substitution on the resonating carbon). Interactions between an incoming group and the carbon framework into which it is introduced have been scarcely investigated (12). However these effects, especially for carbons, may be quite large, as evidenced in iodine substitution with shifts varying from -16.3 to +40.8 ppm (see Figure 4).

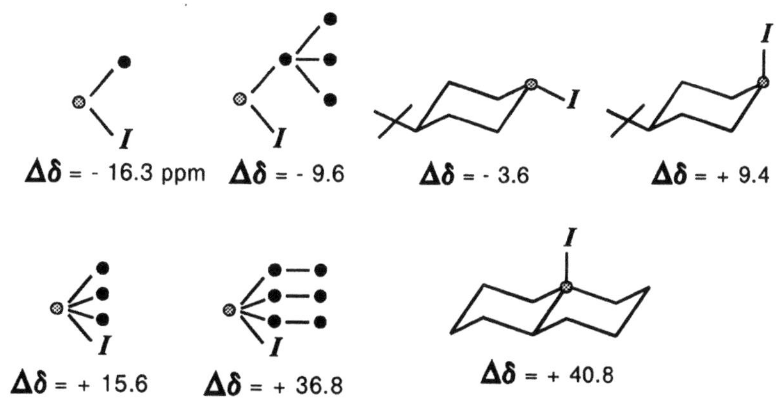

Figure 4. Evolvement of alpha substituent shifts for hydrogen →iodine substitution: indicated values represent the induced shift by substitution on the resonating carbon (grey circle). Dark circles (in acyclic structures) specify the carbon environment.

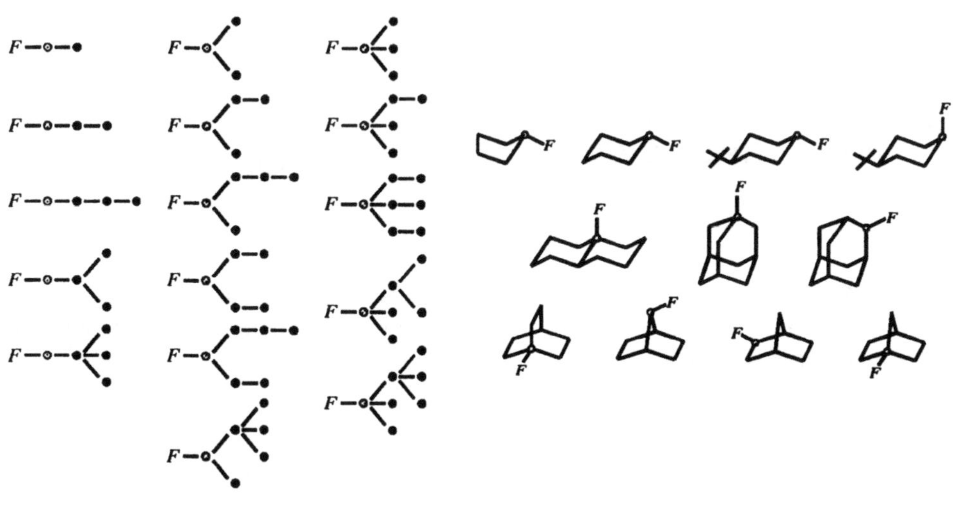

Figure 5. The cyclic and acyclic environments studied. F represents the functional group introduced (F = Me, OH, COOH, F, Cl, Br, I.).

The common assumption of constant substituent increments is therefore definitely inadequate. After a first approximate treatment via factor analysis (12), we tackled this problem using an ANN approach. *a* substituent shifts were analyzed for 16 acyclic and 11 cyclic alkyl environments for seven incoming substituent groups (see Figure 5).

Structural descriptors given as inputs comprise the connectivity of the carbon 13, the size of the cycle (5-, 6- members), the shift of the methylated analog, and for specifying the stereochemistry, the number of B positions "seen" by the incoming group (a notion we already introduced in our previous factor analysis treatment). Figure 6 summarizes results obtained for some of the substituent groups investigated (mean error for the set of re-calculated values and maximum deviation observed). Beyond the fairly good results obtained on account of the large (and generally ignored) variations of these *a* induced shifts, the importance of the seventh input "seen B sites" is quite noticeable.

input parameters with or without B sites

		6	7
HO	mean	0.6	0.7
	max.	2.2	2.6
Cl	mean	1.4	0.7
	max.	6.0	3.6
I	mean	4.6	1.0
	max.	19.5	2.7

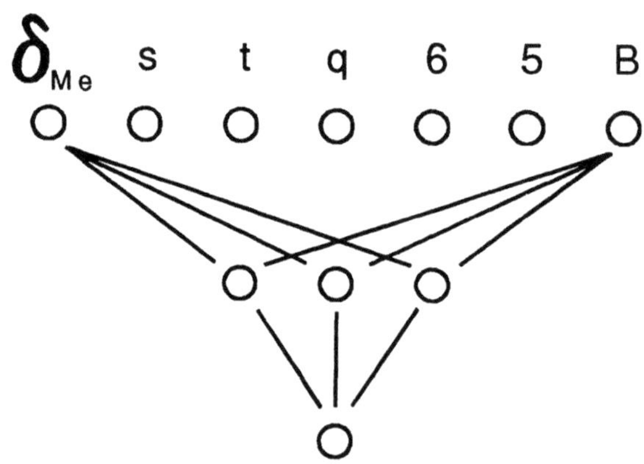

Figure 6. Cross validation results for some substituents : mean, max respectively correspond to the mean error for all the carbon environments studied, and the maximum error observed.

The network comprises six or seven inputs depending on whether the "seen B sites" are (or are not) considered. In the network scheme, δMe is the shift of the methylated analogous; s, t, q describe the connectivity of the ^{13}C; 6, 5 the nature of the cycle (binary encoded), and (possibly) B the number of "seen B sites".

CONCLUSION

As to property prediction (such as ^{13}C shift evaluation) the neuromimetic approach using structural descriptors appears to compete favorably with common empirical models, relying on linear relationships or additive increments.

Better efficiency is obtained, with (particularly) no need to split data into separate subsets. But the main advantage certainly is that there is no need for a pre-defined model. The network learns by itself and constructs its own rules that it can then apply to unknowns.

However, to fully benefit from ANN capabilities some questions remain. Defining the best architecture of the network, selecting the best adjustment algorithms (rapid and avoiding local minima) are still open challenges. But, from the chemist's point of view, the crucial problem to solve is the choice of the best structural representation to input into the network. In other words what information should be given to the network and how should it be expressed?

Keywords

Neural networks, property prediction, ^{13}C NMR, non linear mapping

REFERENCES

[1] a] Zupan J., Gasteiger J., Neural Networks: a new method for solving chemical problems or just a passing phase?" Analytica Chimica acta, 248, 1-30 (1991).
b] Gasteiger J., Zupan J., Neural Networks in Chemistry, Angewandte Chemie, International Edition in English, 32, 503-527 (1993).

[2] a] Rumelhart D.E. and Mc Clelland J.L., Parallel Distributed Processing, Exploration in the microstructure of cognition, vol.1: Foundations, MIT, Cambridge, MA (1986).
b] Mc Clelland J.L., Rumelhart D.E., Explorations in Parallel Distributed Processing: A Handbook of Models, Programs and Exercices, MIT, Cambridge, MA (1988).

[3] Panaye A., Doucet J.P., Fan B.T., Feuilleaubois E., Rahali El Azzouzi S., Artificial Neural Network Simulation of ^{13}C NMR Shifts for Methyl-Substituted Cyclohexanes, Chemometrics and Intelligent Laboratory Systems, 24, 129-135 (1994).

[4] Doucet J.P., Panaye A., Feuilleaubois E., Ladd P., Neural Networks and ^{13}C NMR Shift Prediction., J. Chem. Inf. and Comput. Sci., 33, 320-324 (1993).

[5] Goodacre R., Neal M.J., Keil D.B., Rapid and Quantitative Analysis of the Pyrolysis Mass Spectra of Complex Binary and Tertiary Mixtures Using Multivariate Calibration and Artificial Neural Networks., Anal. Chem., 66, 1070-1085 (1994).

[6] De Weijer A.P., Lucasius C.B., Buydens L., Kateman G., Heuvl H.M., Mannes H., Curve Fitting Using Neural Computation, Anal. Chem., 66, 23-31 (1994).

[7] Wythoff B.J., Levine S.P., Tomellini S.A., Spectral Peak Verification and Recognition Using a Multilayered Neural Network., Anal. Chem., 62, 2702-2709 (1990).

[8] Long J.R., Gregoriou V.G., Gemperline P.J., Spectroscopic Calibration and Quantitation Using Artificial Neural Networks., Anal. Chem., 62, 1791-1797 (1990),.

[9] Blank T.B., Brown S.D., Non linear Multivariate Mapping of Chemical Data Using Feed-Forward Neural Networks., Anal. Chem., 65, 3081 (1993).

10] Weigel U.M., Herges R., Automatic Interpretation of Infrared Spectra: Recognition of Aromatic Substitution Patterns Using Neural Networks., J. Chem. Inf. Comput. Sci., 32, 723-731(1993).

[11] Ricard D., Cachet C., Cabrol-Bass D., Forrest T.P., Neural Network Approach to Structural Feature Recognition from Infrared Spectra, J. Chem. Inf. Comput. Sci., 33, 202-210 (1993).

[12] Doucet J.P., Panaye A., Yuan S.G., Dubois J.-E., Evolvement of Alpha Substituent Shifts in ^{13}C NMR: Application of the DARC PULFO Topological Model for Acyclic Derivatives, J. Chim. Phys., 81, 607-611 (1985).

[13] Dalling D.K., Grant D.M., Carbon 13 magnetic resonance. XXI. Steric Interactions in the Methylcyclohexanes, J. Am. Chem. Soc., 94, 5318-5324 (1972).

[14] Dalling D.K., Grant D.M., Carbon 13 magnetic resonance IX. The methylcyclohaxanes, J. Am. Chem. Soc., 89, 6612-6622 (1967).

[15] Mann G., Kleinpeter E., Werner H., Konformations-untersuchungen mit hilfe der ^{13}C-NMR spektroskopie: III ^{13}C-NMR spektroskopische verschiebungen durch sterisch gehinderte methylgruppen, Organic Magnetic Resonance,11, 561-564(1978).

Clean-up of Spectro-Structural Databases using Artificial Neural Networks.

Claude CACHET[1], Christophe CLEVA[1], Ahmad EGHBALDAR[1], Thierry LAIDBOEUR[1], Daniel CABROL-BASS[1] and Thomas P. FORREST[2]

[1] LARTIC, Université de Nice Sophia-Antipolis, F 06108 Nice Cedex, France
[2] Department of Chemistry, Dalhousie University, Halifax, Nova Scotia, B3H 4J3 Canada

Abstract

Quality control in the construction and usage of factual databases is a well known problem. Classical methods of data inspection are not adequate for spectro-structural data of a large number of molecular spectra of various origins (IR, MS, NMR, etc.). Artificial neural networks can be successfully used as non-linear mapping devices between spectroscopic and structural features. We have built and trained hierarchical neural networks to recognize the presence or absence of several functional groups in a molecule from its infrared spectra. Thanks to the speed of computation of these neural networks, it is possible to scan a large spectro-structural database in order to identify doubtful spectra and/or structures. Typically, these represent only two to 15% of the records.

Resume

Le contrôle de la qualité des données est nécessaire pour les producteurs et pour les utilisateurs des bases de données factuelles. Les méthodes conventionnelles d'inspection des données par des spécialistes ne conviennent pas pour les données spectro-structurales d'origines diverses (IR, SM, RMN etc.). Les réseaux neuro-mimétiques sont utiles car ils permettent des associations non linéaires entre des caractéristiques spectroscopiques et des éléments structuraux. Nous avons développé et effectué l'apprentissage de réseaux neuro-mimétiques pour détecter des groupes fonctionnels moléculaires à partir de leur spectre infrarouge. La vitesse de traitement d'un tel réseau permet alors d'examiner rapidement l'intégralité d'une base de données et d'identifier les couples «spectre - structure» de mauvaise qualité. Ces cas constituent seulement entre deux à 15% des enregistrements.

1 Introduction

The prime importance of quality control in the construction and usage of factual databases is widely acknowledged. Classical methods of data inspection by specialists are not adequate in the case of large databases combining very complex data with high information content. This is precisely the case of spectro-structural databases in which molecular spectra of various origins (IR, MS , NMR, etc.) are associated with chemical structures. Thus an automatic method to alleviate the burden of manual data validation is highly desirable. In this contribution we show that multi layered artificial neural networks can be trained to identify doubtful structures or spectra of poor quality in the case of infrared spectroscopy. The method could be generalized to other type of factual databases.

1.1 Principle

During the last few years, several independent groups have shown that multi layered artificial neural networks (ANN) can be successfully used to recognize the presence or absence of structural features (SF) in unknown molecules from their infrared spectra [1-6]. Typically, in such an approach each individual spectra is coded to make up an input vector ***Iv*** whose elements are usually real values. The relevant SF of the corresponding chemical structure are also coded to make up an output vector ***Ov*** whose elements are Boolean values usually represented by 1 or 0. A training set is selected from the database and the pairs ($\boldsymbol{Iv}_i$,$\boldsymbol{Ov}_i$) of vectors are repeatedly submitted to the network in order to adjust the weights of the connections between the processing units. While this training phase might be time consuming and requires caution to avoid over-training and other ill-behavior of the network, the subsequent usage of trained networks is straightforward and very fast. In favorable cases, the level of performance of artificial neural networks can reach or even outdo the performance of human specialists [6]. However this conclusion is controversial [7].

In the course of developing a decision support system for structure elucidation from spectroscopic data we are building-up a hierarchical neural network (see Figure 1) using infrared spectra as input.

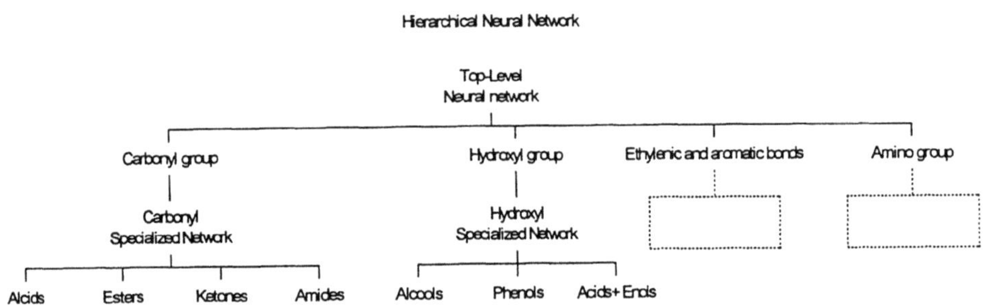

Figure 1.

The top-level neural network is charged with recognizing four very general SF, namely : 1/ carbonyl group 2/ hydroxyl group 3/ ethylenic and aromatic bonds 4/amino groups (excluding tertiary amines or amides). Specialized neural networks are activated when one or several outputs of the top-level network indicate the presence of these groups in the molecular structure. Currently only the specialized networks for carbonyl and hydroxyl groups have been built and trained to recognize the following more specific SF: 1-1/ carboxylic acids 1-2/ esters 1-3/ ketones 1-4/ amides (excluding tertiary amides) and 2-1/ alcohols 2-2/ phenols 2-3/ acids and enols.

1.2 Method

We used three layered feed-forward networks constructed with the ASPIRIN software[1] running on a IBM RS6000 workstation. Three sets of molecules with their infrared spectra are extracted from a collection of 10 000 provided us by Bio-Rad Research Laboratories[2] : a training set used during the training phase, a thresholding set used to determine the threshold values (see infra), and a testing set used to evaluate the network's performances in generalizing to molecules not included in the training set. For each network, the size of these sets are indicated in Tables 1-3. The input vectors are constructed by digitalization of the infrared spectra with a 4 cm^{-1} and 12 cm^{-1} resolution. The results were found to be almost insensitive to a reduction of the resolution to 12 cm^{-1}. The same conclusion was pointed out recently in ref. 7. The output vectors are built automatically by an expert system (SERSS) written in Prolog which recognizes the functional groups from the structure files.

The classical backpropagation algorithm is used to adjust the connection weights during the training phase with the following parameters : learning rate = 0.2, inertia=0.95. To avoid over-training, the network's behavior is continuously monitored by recording the indices of performance on the training (recognition) and on the testing (generalization) sets. The optimal state with regard to generalization performance is then retained. In order to reduce the size of the networks the number of processing units in the hidden layer is decreased as long as the performances are not deteriorated. Moreover, for the top-level network, the size of the input network was further reduced by removing ranges of data points which contribute very little to the output. Thus only data in the ranges 500-1820 cm^{-1}, 2192-2228 cm^{-1} and 2472-3608 cm^{-1} are retained. For the specialized networks the whole range between 500-3700

cm^{-1} is kept, because it was expected that these networks could make use of information different from that of the top-level network. In order to use the network for recognition of structural feature N° i (SFi), the corresponding output value Out_i (a real number between 0 and 1) must be converted to a Boolean value. This is achieved by setting for each output two threshold levels : AL_i (Acceptance Level) and RL_i

[1] ASPIRIN V5.0. The MITRE Corporation. 7525 Colshire Drive, McLean, VA 91024, USA

[2] Copyright © 1980-1993 Bio-Rad Laboratories, Inc. Sadtler Division, 3316 Spring Garden Street, Philadelphia, PA 19104, USA. All Rights reserved.

(Rejection Level), so that the SF*i* is said to be absent if $Out_i<RL_i$, present if $AL_i<O_i$ and undetermined if $RL_i<=O<=AL_i$. The values of AL_i and Rl_i are determined by minimizing a cost function calculated as the sum of the correct minus the erroneous answers.

1.3 Results

The architectures of the networks, number of spectra in the sets and performances are collected in the following tables.

Table 1. Architecture and performance of the top-level network.

Architecture 202x28x4	Training set = 750	Threshold = 250	Test set = 1108

Output	% present	Qr	Pf	Af	EQr	DI
Carbonyl group	45.6	96.5	95.7	92.2	85.3	4.51
Hydroxyl group	26.9	93.0	72.8	92.0	66.5	2.82
Double and aromatic bonds	72.1	88.3	91.0	56.6	53.8	1.94
Amino group	22.7	93.2	58.6	93.2	58.3	2.70

Table 2. Architecture and performance of the carbonyl specialized network.

Architecture 267x35x4	Training set = 750	Threshold = 250	Test set = 493

Output	% present	Qr	Pf	Af	EQr	DI
Carboxylic acids	20.9	97.7	95.1	94.9	84.7	5.41
Esters	31,0	95.1	78.4	89.4	67.3	3.37
Ketones	20,1	90.0	42.4	97.2	57.1	2.32
Amides (I or II)	20,9	96.5	74.8	94.6	71.2	4.16

Table 3. Architecture and performance of the hydroxyl specialized network.

Architecture 267x30x3	Training set = 750	Threshold = 250	Test set = 1108

Output	% present	Qr	Pf	Af	EQr	DI
Alcohols	32.1	94.6	97.2	91.5	84.7	4.23
Phenols	23.6	90.6	80.9	91.5	69.6	2.76
Carboxylic acids + Enols	42.9	93.4	95.3	90.0	84.2	2.64

The indices of performances presented in Tables 1-3 are : Qr : Quality of responses = ratio of the number of correct responses to the number of responses given excluding undetermined cases; Pf : Fraction of Present found = fraction of those compounds containing the SF that were correctly retrieved; Af : Fraction of Absent found = fraction of those compounds that were correctly retrieved in which the SF was absent; EQr : Extra statistical quality = is a measure of the difference between the observed quality of responses and that which would have been expected from a priori probability only; and DI = Discrimination index = gives an indication of the network's ability to discriminate between molecules in which the SF is present and those in which it is absent. For a mathematical definition and discussion of these indices see ref. 8.

These results show that the ANN is able to associate the presence or absence of the SF in the molecule with the infrared spectra with a fairly high degree of reliability even for molecules which were not included in the training sets.

2 APPLICATION

In a second step the whole database of 8620 molecules was scanned with the top-level ANN. For each molecule the output vector computed by the ANN from the infrared spectra was compared to the target vector deduced by the SERSS expert system from the molecular structure stored in the database. A « severe discrepancy » or « mismatch » was counted each time the absolute difference between output and target values exceeded the threshold 0.9. The number of cases for which one or several « mismatches » was observed are reported in Table 4.

Table 4. Results on the whole database

Number of mismatches	Number of doubtful records	Percentage of doubtful records
1	954	11.07 %
2	110	1.27 %
3	8	0.09 %

These figures show the great reduction of the number of cases to be individually validated by specialists. Furthermore, application of the specialized networks to these « doubtful » records leads in many cases to a better agreement between the output and target vectors. Close examination of the « doubtful » cases shows that, in databases of good quality such as the one we are using, most of the discrepancies originate from structural specificity of the molecule, for example, presence of tantomeric forms recognized by the network on the spectra and not explicitly represented in the recorded structure. Crude errors in structures coding, such as an omission in connectivity of the molecular graph, wrong atom valences, etc.; which we have deliberately introduced are readily detected by the SERSS expert system during the target vector generation.

Questionable structures resulting from a particular representation of the mesomeric forms are generally identified by more than one mismatch. The network capability to detect spectra of poor quality is much more difficult to evaluate. For this purpose, experiments are in progress by progressively distorting the original spectra.

CONCLUSION

Although the construction and training of ANN might be computer time demanding, application of trained ANN to a large collection of spectra-structure pairs is very fast and leads to the identification of «doubtful» records. The great reduction in the number of cases to be examined alleviates the burden of database validation by manual inspection. This approach could probably be applied to other types of factual database to identify «abnormal» cases.

KEYWORDS

Data validation, spectroscopic databases, structure, artificial neural networks.

REFERENCES

[1] Robb E.W., Munk M.E., Mikrochim. Acta (Wien) , 1, 131-155 (1990).
[2] Munk M.E, Madison M.S., Robb E.W., Mikrochim. Acta (Wien), 2, 505-514 (1991).
[3] Meyer M., Weigelt T., Anal. Chim. Acta, 265, 183-190 (1992).
[4] Meyer M., Meyer K., Hobert H., Anal. Chim. Acta, 282,407-415 (1993).
[5] Ricard D., Cachet C., Cabrol-Bass D., Forrest T.P., J. Chem. Inf. Comput. Sci., 33, 202-210 (1993).
[6] Smits J.R.M., Schoenmakers P., Stehmann Sijstermans F., Kateman G, Chemom. Intell. Lab. Sys., 18, 27-39,(1993).
[7] Visser T., Luinge H.J., van der Maas J.H., Anal. Chim. Acta, 296,141-154 (1994).
[8] Sbirrazzuoli N., Cachet C., Cabrol-Bass D., Forrest T.P., Neural Comput & Applic.,1, 229-239 (1993).

Chapter 3

CAD SYSTEM ARCHITECTURE AND EVALUATION CRITERIA

COMPUTER INTEGRATED AND INFORMATION AIDED INVESTIGATION ACTIVITIES

Zbigniew KIERZKOWSKI

Technical University of Poznan, Pl. Sklodowskiej-Curie 5, 60-965 Poznan, Poland

ABSTRACT

We present chosen elements of model synthesis and implementation of computer aided activities. Possibilities, of computer aided enlargement and group problem solving with computer integrated multiusers, are taken into consideration. Among these problem solving activities, many have a creative character — these are determined as investigation activities. Computer aiding of investigation activities is based on computer oriented structuralization of various data and knowledge carried out via a computer integrated information exchange (CIIE). CIIE systems and investigation information aiding are important elements of different variants of new information systems for science, technology and economics, systems presently created on the basis of the general model of computer integrated activities organization (CIAO).

RESUME

Divers éléments d'un modèle de synthèse et d'implémentation d'activités aidées par ordinateur sont présentés. Les potentiels d'assistance élargie par ordinateur et de méthodes de résolution de problèmes par groupes sont pris en considération dans un espace intégré d'utilisateurs nombreux. Au centre de ces activités de résolutions de problèmes,certaines ont un caractère de créativité, traitées comme des recherches classiques. L'assistance par ordinateur d'activités d'investigations est basée sur la structuration informatisée de données variées et de connaissances implémentées également dans un système d'échange intégré d'information (CIIE). Les systèmes CIIE et l'assistance à l'intégration de l'information constituent des éléments importants de différentes variantes des systèmes nouveaux de la science , de la technologie et de l'économie. De tels systèmes sont crées actuellement dans le cadre d'un modèle général d'organisation d'activités intégrées et informatisées (CIAO) .

1 Computer Aided Activities Systems Development and New Concepts

We organize various activities, i.e. research, designed, productive and economic processes, based on practical and laboratory experiments, statistical and critical data evaluation as well as on knowledge structuralization.

Despite the many years computers have been in use, detailed problems of computer aided activities [3] still define evaluation aims in almost all problem domains[1] .

In the beginning various computer aiding systems were created as follows: computer aided design (CAD), computer support management (CSM), computer aided investigation activities (CAIA).

Computer aiding is based on creation of computer aided-ware (computer representation of problem domains - PD) and dialogue (D) in the computer (C) - user (U) interaction. In reference to design [2, 4, 5], design-ware include: data and algorithms (programs) bases, computer catalogues, solution bases, factographic data bases - solution space bases.

Regarding management [7,8], management-wares include: common data files, data bases, searching and service of record-keeping activities procedures. Regarding research [1, 6], computer aiding resources include various data bases for science and technology.

We observe continuous enlargement of aiding activities. To the main factors we can add computer services evolution (Table 1). Aiding routine activities and aiding activities more and more adapted to needs of users, are the result of general use of traditional computer services such as: data processing (computing), data storage, user-

1 There is no methodology development (for research, design, management, ...) which is not computer oriented.

computer interaction. Since the mid 1980's the services of the internetworking and intercomputer connections have grown significantly.

It is possible to configure multistation computer structures for information systems. Computer use in the network environment gives rise to:

- using information resources and the computer integrated information exchange (CIIE) system creation,
- possibilities of creating computer integrated activities organizations (CIAO) systems.

Table 1. The evolution of computer services and their integration in creating information systems.

Hardware and software	Computer services	Dominant aiding range	Computer resources	Information systems classes, types of computer aided resources
COMPUTER AIDING SYSTEMS (CAS) AND COMPUTER AIDED ACTIVITIES				
Computer systems	Data processing (computing), simulation, data storage	Computer aided routine activities	Data bases (banks), interaction collecting and information searching processes	Computer aiding systems - CAS: — computer aided design (CAD), — computer support management (CSM), — computer aided manufacturing (CAM), — computer aided investigation activities (CAIA). Computing-ware, integrated computing-ware, aiding-ware
Dedicated hardware-software structures	As above and user-computer interaction	Computer systems for aiding activities fulfilling qualitative new requirements of users	Factographic data bases: computer oriented information structuralization, collecting and searching information processes management, inference processes	Computer aiding systems with support by humans - H/CAS. Integrated aiding-ware
COMPUTER INTEGRATED ACTIVITIES ORGANIZATION (CIAO) SYSTEMS				
Dedicated and open multiuser computer structures	As above and internetworking and intercomputer connections: (a) communication	**Computer integrated and information aided investigation activities**	Information resources: computer oriented data and knowledge structuralization	Computer integrated activities organization — CIAO: A — computer integrated information exchange (CIIE), B — computer integrated manufacturing (CIM),
	(b) application	Computer aid to problem solving with the participation of group users	Information resources group creation and exchange co-ordination	C — computer integrated technical and economic management, D — computer integrated administration. Group-ware

The CIAO is based (Figure 1) on:

- Using application and communication internetworking (local or widespread when necessary) in the multiuser environment (C0, C1, C2, ...),
- Separation of group problem solving in the user environment (U0, U1, U2,...)
- Synthesis of computer aided resources, as in solutions for creating as C1-U1, C2-U2, ... required in problem solving, synthesis will be completed with programming realization of activity integration co-ordination (C0-U0) by using internetworking services.

There are many issues. The requirements for formulation from the start of activities methodology in detailed computer oriented problem domains, based on computer data and knowledge and their exchange are presented.

Besides the new information systems we distinguish the following classes:

A Computer integrated information exchange,

B Computer integrated manufacturing combining design and management activities aiding in manufacturing systems,

C Computer integrated technical and economic management in organizations,

D Computer integrated administration.

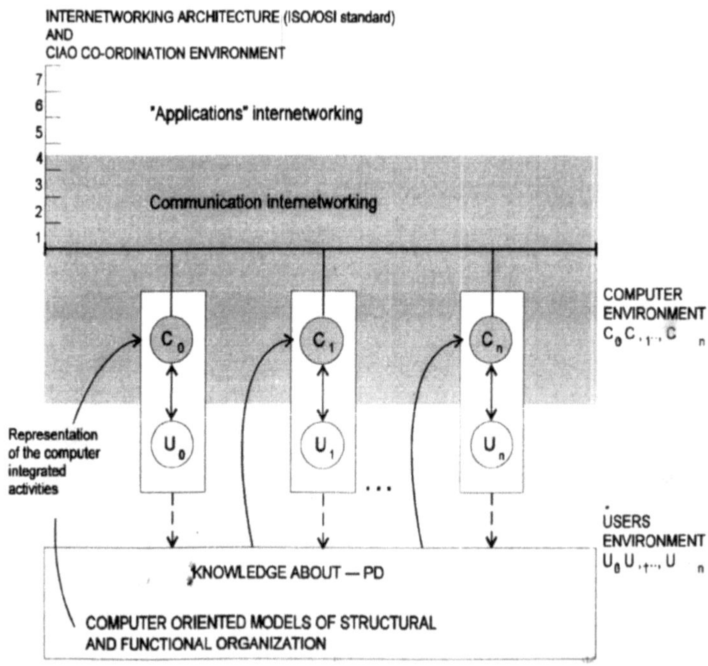

Figure 1. Multiuser structures in the network environment for the computer integrated activities organization (CIAO).

In all cases of information systems, the CIAO is based on information aiding of activities and the CIIE. It leads to posing issues of problem solving within composed organizations with group participation of multiusers.

2 AIDING INVESTIGATION ACTIVITIES AND COMPUTER INTEGRATED INFORMATION EXCHANGE

Improvement of activities through their computer integration provides possibilities for problem solving. In problem solving we can separate activities of a creative character. These are determined by investigation. Such activities dominate in construction and technological design and in research. Computer aided investigation activities are based on facilities of various data and knowledge. In general computer use working in the network, cooperative (group) storing and information exchange is possible. CIIE systems are factors of new solutions of structural and functional CIAO and new information system models (for research, manufacturing, management etc.).

Development of the scope of aid is visible in succeeding generations of information systems. At the start it includes information aid of activities for concurrent statistical conclusions. Next we build and use data banks for science and technology, and with information search procedures for activities aiding, improving critical evaluation of experimental data. Further expansion of the range of activities aiding includes improvement of inference processes on the basis of experience and both general and detailed knowledge accessible for aiding. Information resources for computer aided information investigation activities include: bibliography, scientific and technical information, text information computer files, archives.

Development of aiding range connects previous solution functions with functions of succeeding generations of information aiding of investigation activities. The main role is played by the creation of computer information resources (see Table 1).

Computer information resources determine the output base of the creation of first CIIE solutions. These solutions can be used for the improvement of practice aiding of investigation activities. Qualitatively - new factors of synthesis (creation) of information aiding systems of investigations are as follows:

- Separation of group problem solving and creation of computer information resources in a multiuser environment,
- Separation of information process logistics in investigations for information aiding co-ordination modeling of group activities,
- Creation of computer integrated information resource exchange systems.

Protection of system functioning of the CIIE is modeled by the logistic structure of information processes. The model of process logistics of the computer integrated and information aided investigation activities will be the graph shown in Figure 2. Nodes correspond to functions of participants' investigation activities. The graph itself is the representation of process logistics of information aiding of investigation activities. In understanding such logistics (see also [7, 9], the edges should be interpreted as information processes between participants of the CIIE system. Then the edge interpretation is as follows: (1) formulating user information needs, (2) information facilities, (3) information needs locally serviced, (4) transferring information from U to IR, (5) information applications directed to IR, (6) using IR.

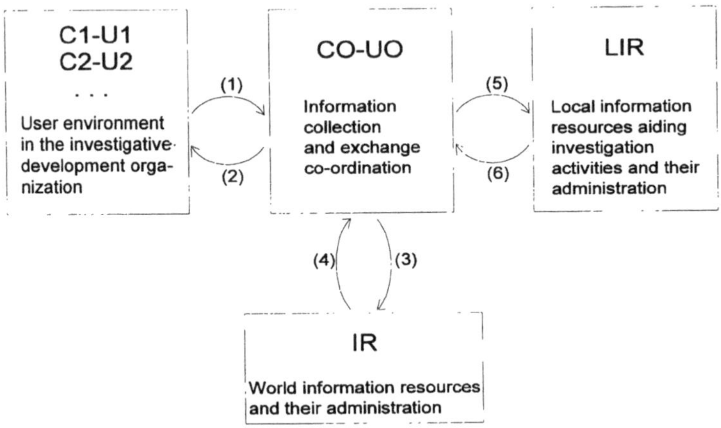

Figure 2. General logistic structure of aiding processes of activities in the CIIE.

In particular CIAO systems for investigations, process logistics realization makes possible the system supplement for the additional element, i.e. "logistic" base (LB). The logistic base LB is the new information resource of the system which is equally maintained and managed in the multistation computer structure, on the K0-U0 station (see Figure 1).

3 Investigation Information Aiding and Computer Integrated Information Exchange

The network environment of computers (a lot of computer station — CS) is to-day the information aiding base of investigations and creates various variants of the CIAO systems.

The cooperation of computer multistations is based on internetworking and intercomputer connections; moreover in the information exchange we use network and application communication services.

Data and knowledge structuralization in network information resources and means for their dissemination.

Among information resources we distinguish the following different types in Table 2.

Table 2. Information structuralization in the network environment and dissemination modes

Information resources	Bibliography	Data for science and technology	Text information	Computer files	Archives
Computer oriented structuraliz.	Catalogues Texts Hypertexts	Data bases of: - facts, - critically evaluated data, - recognition patterns, - inferences interpretation patterns	Letters Documents	Software Text information Graphic information Sound information	Archives including computer files
Network services modes	telnet gopher	telnet "PAD terminal access"	e-mail telnet gopher news	telnet + ftp gopher	telnet ftp gopher ... ...
Resource storage	Centrally	Centrally	Distributed	Distributed	Distributed
User participation	Using	Using	Using and transferring	Using and transferring	Using and transferring

A central utilization of information and distributed information resources are considered within such an approach.

Dissemination of local information resources in world networks.

A variety of creation and storage facilities of information and implementations can be observed (Figure 3). The structure of information resources is homogenous.

At first it is visible that independently of an environment for creating information resources (for example, dBASE, ICES, C, Pascal, Progress, Informix, etc.), information resources have a strictly determined structure which includes: management software (programs of EXE type), data (or resource) formats.

Next we see that among information resources we can distinguish:

Information file resources (FR),

Data base systems (DBS).

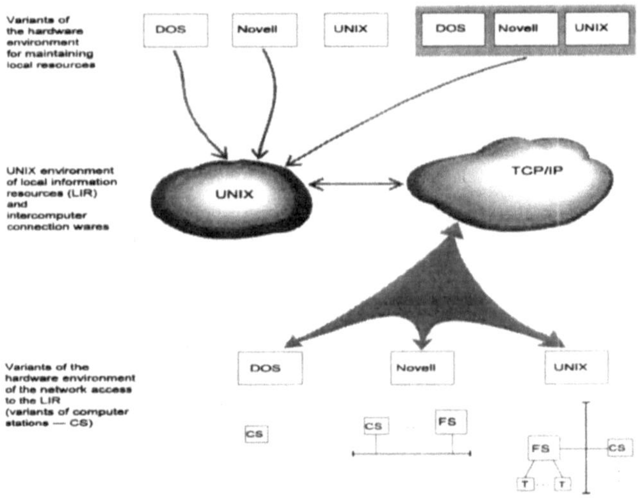

Figure 3. Implementation environment of the network dissemination of the LIR.

Let us note that in information exchange we have various environments of network storage and dissemination of information resources. Variants of hardware environment of resource implementation include: DOS, Novell, DOS-Novell-UNIX. We only utilize UNIX environment to transfer resources within internetworking (for example TCP/IP), and again hardware environment variants of the network access to information resources include: DOS, Novell, UNIX, Windows.

We can recommend the following cases as simpler variants of dissemination of local information resources with creation of CIIE systems:

(1) Dissemination of information file resources (FR). Functional organization of storage and exchange of file resources is based on the service choice and determination of the information application and then - within the chosen service - on the "information exchange", enclosed in the file resource.

(2) Information dissemination within DBS implemented on network servers. Functional organization of storage and information exchange is based on the choice of DBS and then on realization of an "interaction session".

(3) Dissemination of information resources implemented on LAN servers. Functional organization of storage and information exchange is based on utilization of the telnet service and realization of a DOS session in UNIX, for example.

Information aiding coordination organization of activities.

In network information exchange, we use internetworking, intercomputer and communication services (see Figure 1).

Internetworking and intercomputer implementation for realizing application services results from information exchange coordination and refers to information aiding of group activities of multiusers (U0, U1,...) working on computers (C0, C1,...).

Figure 4 shows the structural and functional organization of the work within the backbone computer network utilizing world information resources[2].

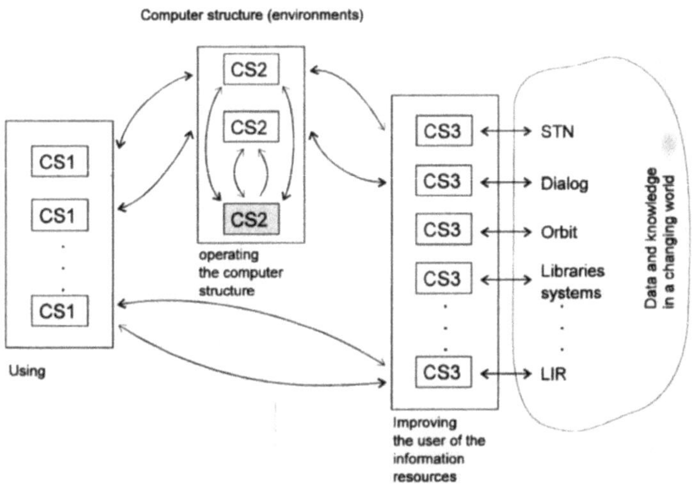

Figure 4. Structural organization of the work within the information computer network exchange (solution experience at Kedzierzyn-Kozle — Poland)

CONCLUSION

Models and synthesis of new information systems created for scientific researches and practice improvement of design, management, manufacturing and administration, are based on the CIAO. Within different classes of information systems, possibilities

2 The solution improving the use of scientific and technical information in the Institute of Heavy Organic Synthesis at Kedzierzyn-Kozle, within LAN using, through WAN, world information resources (i.e. STN, Dialog, Orbit, library systems, common resources). The following notations are accepted: CS1-stations of multiusers, CS2-stations operating the computer structure, CS3-stations for improving the use of information resources. Problems solved with group participation of multiusers are as follows: exchange and transfer of files in aiding investigation activities, information exchange in the sphere of chemical analytics, aiding investigations in the sphere of chosen technologies, aiding marketing activities.

of information aiding of investigation activities on the base of the CIIE, have appeared.

Main factors of the computer integration of information aiding of activities are as follows:

- Creation of computer networks with local, regional, domestic, continental and intercontinental connections,
- Configuring multistation computer structures and use of internetworking and intercomputer connections adapted to activities aiding,
- Use of world information resources,
- Information exchange coordination for group problem solving in new models of information systems.

KEYWORDS

Network, information system, computer aiding systems, information aided activities

REFERENCES

[1] Bylicki A.(ed.), Domestic Conference: Physico-Chemical data bases for science, education, industrial systems simulation - Polish, Zakopane, Proceedings (in preparation) (1993).

[2] Crips R.J., Smith F., An information system to support the design activity, Journal of Engineering Design, 4, 1, 3-10 (1993).

[3] Gasparski W. (ed.), Design studies - The elements knowledge on design - in Polish, WNT, Warszawa (1988).

[4] Kierzkowski Z., The evolution and integration of computer aided design systems. International Conference on Engineering Design, ICED 91, Zurich, Edition HEURISTA, 2, 861-866 (1991).

[5] Kierzkowski Z., Design-ware synthesis. International Conference on Computer Integrated Manufacturing 1992, Politechnika Slaska, Mechanika, series, Z. 108, Gliwice, 189-196 (1992).

[6] Kierzkowski Z., Computer network use and open systems services for investigation information aiding — in Polish, II Domestic Conference on Science and Technical Information, Jastrzêbie Zdrój. Proceedings, 91-106 (1993).

[7] Kierzkowski Z., Manufacturing group-ware synthesis or computer integrated information aiding design and management activities. International Conference on Computer Integrated Manufacturing 1994, Politechnika Slaska, Mechanika, series, Z.117, Gliwice;.321-330 (1994).

[8] Kierzkowski Z., Plonka-Szydlak E., The evolution and integration of management support computer systems, Computer in Industry, Elsevier Science Publishers, Amsterdam, 16,.69-79 (1991).

[9] Korzen Z., The present status and prospects for development of logistic (in Polish). Problemy Maszyn Roboczych. Logistyka. Komitet Budowy Maszyn PAN,. 2, Z. 2, 35-52 (1993).

A Cooperative Design Environment based on an Active Repository

Jessica BARROS de SA[1]

Laboratoire PRISM, Université de Versailles, 45 av. des Etats-Unis, 78000 Versailles, France
E-mail: Jessica.Barros-de-sa@prism.uvsq.fr

Abstract

In a cooperative design environment, a group of designers work together in harmony in order to develop a common product. This environment must encourage the communication between designers according to the team's organization and task distribution. In general, each designer is responsible for a number of tasks, and the synchronization of all tasks is fundamental for a successful production process. The aim of this work is to assure the cooperation between Kheops' users through a repository shared by the design environment. This repository stores information about the agents, the tools, and the design data. Moreover, our repository offers a set of rules in order to ensure the functionality of the cooperation.

Resume

Dans un environnement de conception coopératif, un groupe de concepteurs travaille ensemble afin de développer un produit en commun. Cet environnement doit encourager la communication entre concepteurs selon l'organisation de l'équipe et la distribution des tâches. En général, chaque concepteur est responsable pour un certain nombre de tâches et la synchronisation de toutes les tâches est fondamentale pour aboutir à un bon résultat. L'objectif de ce travail est d'assurer la coopération entre les utilisateurs de Kheops à travers l'utilisation d'un répertoire partagé. Ce répertoire stocke des informations concernant les concepteurs, les outils et les données liées au projet en développement et, en plus, il offre un ensemble de règles afin d'assurer que la coopération sera fonctionnelle.

1 Introduction

A cooperative design environment is characterized by a set of tools which permits a group of designers to work together in order to develop a software system. In general,

---

[1] Supported by CAPES/Brazil

each designer is responsible for a number of tasks, and the synchronization of all tasks is fundamental to a successful software system. This synchronization requires a dynamic iteration between designers and the use of a shared workspace to exchange information. According to [Chak92], cooperative work involves extensive access to information repositories, tools for computation support, decision making and/or inferencing using a set of criteria, and coordination of tasks execution.

The goal of this work is to integrate designers' cooperation into Kheops [Bouz90], a development environment for modeling object oriented database applications. In order to do that, we propose a new functional architecture for Kheops based on the client-server approach. This kind of architecture provides many facilities concerning concurrency control, data integrity, and workspace management.

Nevertheless, it is not sufficient to satisfy all requirements of a design environment. It is necessary to introduce the management of events, the concepts of nested transactions and conception team in order to improve the cooperation and communication between the developers of a given application. In a cooperative design environment, it is important that a designer knows the tasks developed by the others and the results of their works. An object update must be considered as an event which can be interesting to all designers or which can generate another update. Taking these events into account, we will increase the cooperation between developers, and consequently, the productivity and the quality of the product.

In our work, the management of the cooperation is supported by the server through the active rules and we use the Event-Condition-Action (ECA) rules [Daya88] to model the interaction between the users involved in a common creation.

The paper is organized as follows. The second section describes the new functional architecture proposed to Kheops. The third section presents some active rules defined for its repository.

2 Functional Architecture

A cooperative design environment is characterized by the following components: agents, information and activity. The agents are either human or tools. The information is the data concerning the software produced, the organizational information or the events that constitute the dynamic information. In our approach, the activity is implemented by giving to the repository the possibility to react to events by triggering rules. In [Jasp94], an active repository is used to partially automate scheduling and controlling constructing software product activities.

However, this automation is not sufficient to guarantee good designer collaboration. It is necessary to offer a dynamic communication between designers, and further to implement some cooperative design rules to ensure consistent work.

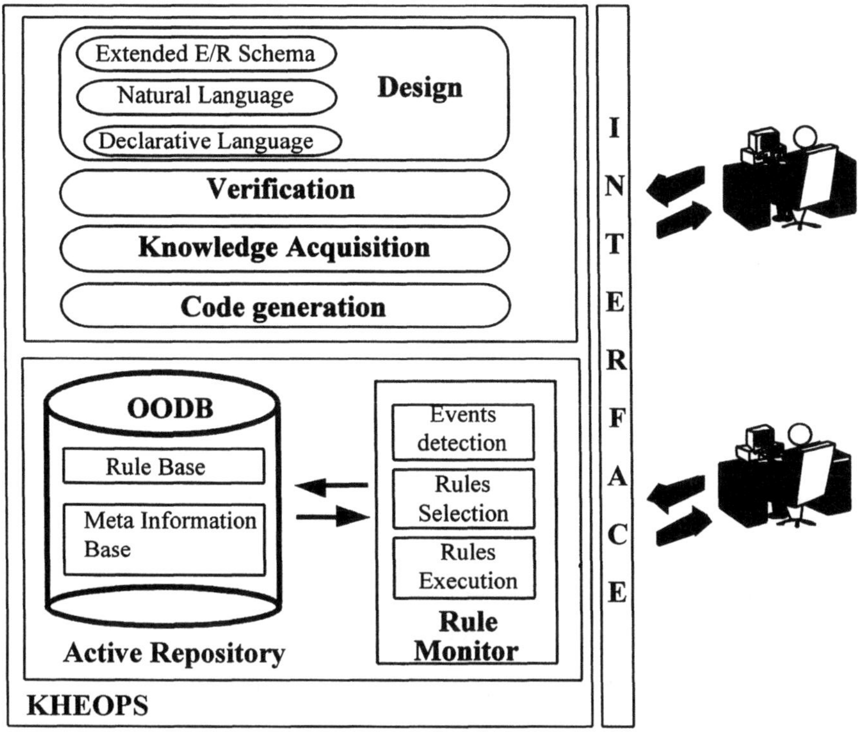

Figure 1. Design Cooperation in Kheops

In order to implement a cooperative design environment, we offer a functional architecture (see Figure 1) based on an active repository. This repository is implemented as a server and the users are clients that share stored data. The developers have an interface through which they can ask for services, such as locking a file, or generating a code. Developers' requests can generate events and these events can trigger rules. When a rule is triggered, its condition is evaluated and if it is satisfied, the corresponding action is performed.

The active repository has two main parts: an object-oriented database and a rule monitor. The database stores the rules and the meta information related to agents and design data. The rule monitor detects the events, and then selects and executes the concerned rules.

3 Active Rules

Using the active rules we assure the management of the activities related to a cooperative design environment. Among others, the rules are used to ensure data integrity, to manage the workspaces, to permit iteration and communication between designers, to respond to agents' requests according to the agent's role, and to coordinate task execution.

We have identified a set of ECA-rules related to the functionality of a cooperative design environment and we offer the designers the possibility to define their own rules. Before integrating a user-defined rule in the repository, the system verifies if it is consistent with the already implemented rules.

A rule is defined by an event, a condition or an action.

An event can be external, based on operation (before/after execution), or time events (absolute, relative to another event, periodic).

A condition is a logical predicate on the database state.

An action can be an operation on the database, a message sent to a developer to enforce cooperation, or a message sent to the tools in order to execute a functionality. An example of a rule to control data coherence is given below:

> **ON** modification by a designer D of the structure of an entity
>
> **IF** this entity was defined by D but used by other designers
>
> **THEN** take the modification into consideration in all D's modules
>
> AND inform the other designers

We have also defined supplementary rules to cover the other two cases: (1) if the entity is used only by D, then take the modification into consideration in all D's modules; (2) if the entity was not defined by D, then send a message to the designer who defined it in order to ask for the modification.

Notice that the active rules are not used only to control data coherence. We give below two examples of rules. Rule 1 is triggered every month and integrates all modules defining a same application, and rule 2 is used to control file manipulation.

Rule 1

> **EVENT** automatic_integration **EVERY** 1 month **WITHIN** [03.31.94 - infinite]
>
> **ON** automatic_integration
>
> **THEN** integrate all modules defining a same application

Rule 2

ON a request to lock a file **IF** the developer is not authorized to lock the file **THEN** send a notification message to the developer

We are researching now how to control the contents of the repository. In our context, the problem is due to the fact that a rule execution can trigger another rule, and the agents can modify the rule base. Consequently, we have to treat the problems of validation, termination, and confluence of the set of rules offered by the repository.

CONCLUSION

The aim of this work is to integrate designers' cooperation into Kheops. Our proposition is based on an active repository. The dynamics of this repository is implemented by the ECA rules. These rules are used to enforce the communication and the cooperation between designers of a given application. The repository offers a set of rules, but the agents can add their own rules.

As future work, we have to improve the cooperative rules in order to permit the immediate visualization of any object modification by all concerned designers. Through this facility, the designers can benefit from all contributions as soon as they arrive.

KEYWORDS

Cooperative design, active databases, ECA rules, repository.

REFERENCES

[Bouz90] Bouzeghoub M.et al, A Design Tool for Object Oriented Databases, Proc. of the 2nd Conf. on Advanced Systems Engineering (CAISE 90),Stockholm, Sweden, May (1990).

[Chak92] Chakravarthy S. et al., Databases: Supported Cooperative Problem Solving, Technical Report UF-CIS-TR-92-046, Department of Computer and Information Sciences, University of Florida (1992).

[Cons92] Constantopoulos P., Doerr M., Vassiliou Y., Repositories for Software Reuse: The Software Information Base, Proc. of European Research Consortium for Informatics and Mathematics Workshop, Methods and Tools for Software Reuse, Heraklion, Oct. (1992).

[Daya88] Dayal U. et al., The HiPAC Project: Combining Active Databases and Timing Constraints, SIGMOD RECORD, 17, 1, March (1988).

[Goda93] Godart C., COO: A transaction model to support COOperating software developers COOrdination, 4th European Software Engineering Conference, Garmisch, LNCS 717 (1993).

[Jasp94] Jasper H., Active Databases for Active Repositories, Proc. of the 4th Int. Workshop on Research Issues in Data Engineering, Houston, Texas, 375-384, Feb. (1994).

[Wido91] Widom J., Cochrane R., Lindsay B., Implementing Set-Oriented Production Rules as an Extension to Starburst, Proc. of the 17th Int. Conf. on VLDB, Barcelona, Spain, Sept.(1991).

[Wiil93] Wiil U.K., Experiences with HyperBase: A Hypertext Database Supporting Collaborative Work, SIGMOD RECORD,.22, 4, Dec.(1993).

A REUSE-BASED SCHEMA CONSTRUCTION SYSTEM

Ana Paula AMBROSIO≠

Laboratoire PRISM, 45 av. des États Unis, 78000 Versailles, France
E-mail: ambrosio@prism.uvsq.fr

ABSTRACT

In an information processing system, the conceptual schema is considered to define the degree of visibility of the context's different elements, as well as their organization and semantics. Therefore, the reuse of a conceptual schema, or parts of it, by several systems would imply the adoption of a common vision as provided by the imported parts. In order for such parts to satisfy the new design requirements, the applied reuse mechanism must be capable of capturing the scheme's semantics. This paper deals with the reuse problems related to conceptual schema construction and the respective semantic requirements. It proposes a retrieval-based system handling imprecise queries for the reuse of schema components.

RESUME

Dans un système d'information, le schéma conceptuel sert a définir le dégrée de visibilité que l'application a des différents éléments du contexte modélisé, ainsi que leur organisation et leur sémantique. Par conséquent, la réutilisation d'un schéma conceptuel, ou des parties d'un schéma, par plusieurs systèmes signifie l'adoption d'un même point de vue. Pour que ces parties satisfassent les nouveaux besoins, le mécanisme de réutilisation doit être capable de capter la sémantique des schémas. Cet article traite les problèmes de réutilisation liés à la construction de schémas conceptuels et les besoins sémantiques correspondants. Il propose un système de sélection qui permet le traitement de requêtes imprécises.

1 INTRODUCTION

The intention of reuse is to allow developers to benefit from already written and validated conceptual schemes (CSs), all the while conforming to the new design requirements. Although existing modeling techniques and tools readily lend themselves to such requirement, they are not enough to promote large scale reuse. The main difficulties met during the exploitation of existing schema definitions are

---

≠ *Supported by CNPq, Brazil*

firstly to discover the suitable elements liable to satisfy the design requirements and then to combine the selected elements in order to form the final schema.

Attempts to retrieve schema definitions solely on the basis of syntactical or structural equivalence (same name, attributes, etc.) are not acceptable, since they do not guarantee conformity to the particular view of the new Universe of Discourse (UoD). Instead, the search for reusable parts must concentrate upon the required semantics in order to ensure consistency between the conceptual schema and the new applications' semantics.

Reusable components are schemata which are catalogued and saved in a repository. Their semantics is captured by analyzing the key concepts that each schema bears. These are then expressed using a set of terms belonging to a defined context. A Term dictionary is used as the reference to the meaning of the different terms and provides the measure for their semantic equivalence. During reuse, components are retrieved by searching the repository for schemata with characteristics that approach the user's request within a given distance.

2 SCHEMA CONSTRUCTION THROUGH REUSE

A CS provides the applications' knowledge about their context, i.e. it defines the degree of visibility of the context's different elements, as well as their organization and semantics. In fact a CS also provides events, operations and constraints (temporal and dynamic) in order to express the context behaviour.

The ANSI organization defines three phases for the design of database systems which are now generally accepted as the basis to the classical approach [ANSI 75]. The first consists in the specification of a conceptual description of the modelled application (i.e. the design of the CS). The next phase concerns the transformation of the CS into a logical schema depending on the data model supported by the DBMS. Finally, the physical design phase defines the physical storage for data and their access methods.

In practice, mapping algorithms exist for the last two phases, whereas CS design remains mainly an expertise activity. Today, the transformation of the CS into a logical schema and the resulting implementation are obtained automatically (semi-automatically) by means of code generators. Therefore, applying reuse would be most significant in the first phase of database systems design. This is directly concerned by the modelling process that leads from an analysis of the application domain to a formal (or semi-formal) representation of said domain. In contrast to the classical design approach where a schema is constructed directly from the design requirements, reuse implies that the designer must endeavor to **construct** a CS mainly from existing elements.

His task is first to investigate and to select those elements that best match his modeling purposes according to the user's requirements. Then, he will combine the selected building blocks, adding his own alterations and modifications, in order to obtain the final schema. Figure 1 shows a Schema Construction System (SCS) based on a repository of schema components.

Knowledge Acquisition represents the process of filling the schema Repository with the reusable components. This is done by applying domain analysis to Source Schemes (already existing ones) in order to identify the reusable components or through the direct analysis of the different domains of application. The components are then catalogued and added to the repository.

During reuse, the Retrieval Mechanism searches the Repository for the components that best match the designer requests. A user-assisted Component Manipulation is then applied to the selected components in order to merge and restructure them into the Target Schema. In order to better understand such a system the reuse-based approach is considered in relation with that of schema integration. Schema integration is used to "combine" two (or more) existing schemes into a global, unified one, all the while keeping the semantic integrity and consistency of the resulting schema.

The process of schema integration consists of the five following iterative steps: canonization, comparison, conciliation, merging and restructuring [Song 94]. It is noted that, in the case of integration, the two source schemes already exist and are taken as input. On the other hand, in a Schema Construction System (SCS), the first schema would be that of the repository and the second one might be considered as a skeleton schema which is described by the designer's requests. Based on this parallelism, we may proceed to consider the required activities for schema construction.

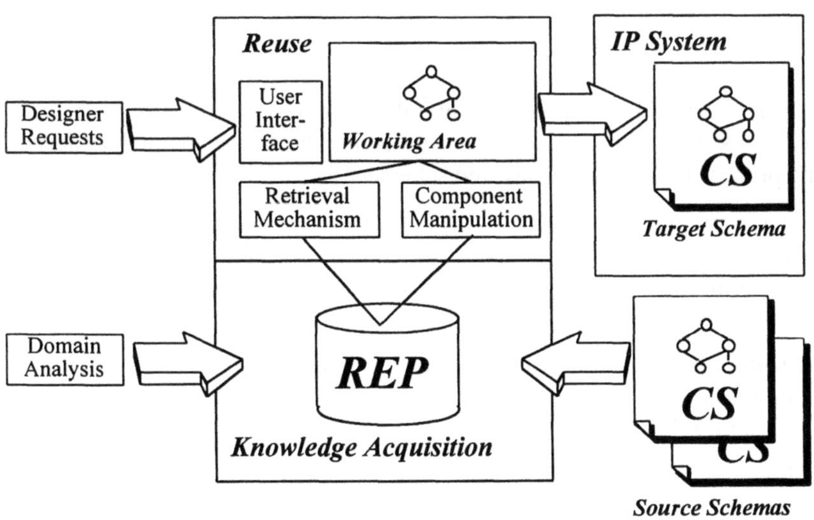

Figure 1. Schema Construction System

The canonizing step is needed in schema integration in order to normalize the intra-schema information and to resolve any detected conflicts. In a SCS, this activity is only required when adding new elements to the repository. Canonizing in schema construction is an off-line activity applied during the knowledge acquisition phase in order to maintain the coherence of the repository.

The next step in schema integration is to associate the semantically equivalent objects belonging to the different input schemes. A pairwise comparison of the schema's objects is first performed to identify objects that exhibit similar semantics followed by a user-assisted conciliation in order to resolve eventual conflicts. In a SCS, this is performed by the query mechanism whose task is to retrieve the components that exhibit semantics similar to the designer request.

It is noted that, whereas comparison involves an analysis of the input schemes in order to capture their objects' semantics, the repository components semantics would be already captured during knowledge acquisition. The automation of such a process is not of prime interest for schema construction.

Finally, in both cases the target schema is constructed. The merging step generates a new schema integrating the components which result from the previous object identification process (either by comparison/conciliation or by retrieval). As a last step, the integrated schema is restructured into a consistent target schema.

3 SCHEMA COMPONENTS AND THEIR CLASSIFICATION

Based on the abstraction/decomposition approach, the CS building blocks are modular sub-schemes defined in terms of components that are stored in a repository and classified according to semantic criteria. Concepts presented in the CS at a given level can in their turn be represented by other components (schemes) to describe the required properties of a concept in more detail. The above method is derived from existing propositions for multi-layered schemes.

Components are thus comparable to the concept of molecular objects [Bato 84] in the relational model, to the "subject and information areas" [Feld 86] in the ER model, and to the complex object type [Kim 87] and the schema object type [Troy 91][Troy 93] in the object oriented approach.

Components are thus schemes that are perceived as a unit with additional information attached, and can be used in the CS as an abstraction of an underlying schema [Ambr 94].

The following template shows the set of attributes that characterize a repository component by capturing its semantic, syntactic and pragmatic aspects. To describe the associated semantics, two properties are defined per component, namely its functional area and its setting, collectively specified by a domain attribute. Components' domain attributes are a set of terms taken from a controlled *facets vocabulary* [Prie 87]. This avoids duplicate and ambiguous descriptors. A term thesaurus groups all the synonyms under a single concept. The term that best describes the concept is selected as the *representative* term that will be included into the set of accepted terms.

Each component is given a name (label) and an informal description intended to briefly explain the semantics of the contained schema. It is possible to find several components, in different hierarchical paths, with the same name. Nevertheless, each component is attributed a unique identifier. Each component has a list of key-concepts that represent the UoD's fundamental properties pictured in the contained schema.

TYPE Component ::=	
id: INTEGER,	* Unique identifier attributed within the repository
label: TEXT,	* Descriptive name given to the component
description: TEXT,	* Natural language descr. of component's contents
domain (	* The component's original domain of application
funct-area: SET Function-Area,	* Describes application dependent activities
setting: SET Setting),	* Describes where the application is exercised
key-concepts: SET Concept,	* List of most important concepts in the schema
schema: Schema	* Pointer to the schema describing the component

The need for abstracting the important concepts from the schema comes from the fact that the same concepts can be interpreted (and therefore represented) differently by different designers (e.g. as an entity or a relationship in the ER model). This is due not only to the vision the schema designer has of the UoD, but mostly to his personal technique of modeling. Others have already identified this handicap which they attribute to the semantic approach based on methodologies [Gust 94].

4 SEMANTIC EQUIVALENCE

During the CS construction, components that best match the required semantics are searched for by means of a retrieval mechanism that allows for solutions with approximate semantics. Approximate solutions are sought in case an exact one is not found, but may also be desirable in order to obtain a wider choice. Components of equivalent semantics are obtained by the relaxation of some characteristic condition(s), e.g. by looking for a key concept synonym instead of an exact term. A query processor is in charge of the automatic modification of semantic conditions by navigating through conceptual graphs describing the components' characteristics (e.g. key concept or label equivalents).

These graphs are contained in two auxiliary dictionaries: A Domain Dictionary defining the relationships between different application domains, and a Term Dictionary defining the semantic and syntactic relationships between concepts. In the Dictionaries, nodes correspond to concepts, and arcs correspond to relationships between these concepts.

The relationships currently accepted in the dictionaries are: Synonym, Similarity, Is-a, Can-be, Part-of and Composed-of. Note that the relationships represented by the first two arcs are reflexive in the sense that if a concept A is a synonym of concept B then concept B is a synonym of concept A (equally true for the similarity relation).

As for the other relations, the Is-a relation is the inverse of the Can-be relation (if concept A is-a concept B, then concept B can-be concept A), and the Part-of relation is the inverse of the Composed-of relation. The distance between (minimum number of arcs of same type connecting) two concepts is used as a measure of the semantic equivalence. The departure point is a condition originally stated in the query (corresponding to a concept node).

Based on the liberties given by the user through the statement of the relaxation conditions to be applied to the original query, the query processor will traverse one arc at a time. The maximum distance to be partaken by the processor can also be defined through the use of a radius statement. If no radius is given, the relaxation condition will be applied only once (i.e. radius = 1).

5 QUERY PROCESSING

In order to allow the relaxation of semantics conditions, the user requests are formulated as imprecise queries using the SELECT statement in an SQL-like syntax. The command structures are adapted to the retrieval of conceptual schema components through the definition of some characteristic condition (*cond*) and to permit imprecise querying using some relaxation criteria (*rlx*) which can be forced *(!)*, based on a given radius (*rad*) :

SELECT ([id],[label],[desc],[key-concepts],[funct-area],[setting])

WHERE ([(id,*cond*,[rlx])],[(**label,***cond*,[rlx])],[(**key-concepts,***cond*,[rlx])])

DOMAIN ([(funct-area, *cond*, [rlx])], [(**setting,** *cond*, [rlx])])

where rlx ::=

([(Syn[!],*rad*)],[(**Is-a[!],***rad*)],[(**Can-be[!],***rad*)],[(**Part-of[!],***rad*)],[(**Comp-of[!],***rad*)])

When stating a query, the user is permitted to state an original condition which he wishes satisfied. This approach corresponds to the traditional querying mechanism used in databases. In addition he can state relaxation conditions to be applied to the original query. These relaxation conditions can be "forced" in the sense that they will be applied to the original query independently of having or not found a component that satisfies the query, or "deferred" which will only be applied if no answer to the original query was found. In this last case, the relaxation conditions are applied one at a time until an answer is found.

The relaxation conditions permitted by the query mechanism correspond to the relations representing the arcs in the conceptual graphs of the dictionaries. This means that when the user states that the original condition can be relaxed by a given relation, the query modification mechanism will traverse the arcs in the conceptual graph that correspond to the relation stated and that have as departure points the concept in the original query.

The new concepts that are reached by this process are then taken as conditions for the newly generated queries. This process can be repeatedly applied, not only departing from the original concept, but from the recovered concepts as well.

CONCLUSION

This paper proposes a new semantic-based approach for conceptual schema reuse, and defines a set of activities for schema construction. The adopted approach is based upon a semantic classification of schema components and an associated retrieval mechanism.

Reusable components are canonized schemata which are stored in a repository and classified according to a characteristic set that represents their associated semantics. A retrieval mechanism which permits imprecise SQL-like queries on the components' classification aspects is defined. The query processor searches for approximate solutions by traversing the conceptual graphs contained in the dictionaries in an ordered manner.

The results obtained are encouraging, and we feel that the development of application conceptual schemes through reuse is a big step towards the production of reliable database applications quickly and economically. Much work is still needed for the automation of the classification process and the refinement of the semantic equivalence evaluation. These aspects are the subject of several research projects related to different domains of conceptual modeling (e.g. schema integration and multi-layered modeling). It can be envisaged that the problems hindering reuse will be surmounted, as reuse still seems to be the solution to the cost-effectiveness problem in the development of large software systems.

KEYWORDS

Reuse, conceptual modeling, selection mechanism, semantic retrieval, imprecise querying

REFERENCES

[Ambr 94] Ambrosio A.P., Introducing Semantics in Conceptual Schema Reuse, CIKM'94, U.S.A., Nov. (1994).

[ANSI 75] ANSI/X3/SPARC Study Group on Database Management Report, Tsichitzis (ed.) Feb.(1975).

[Bato 84] Batory D.S., Buchmann A.P., Molecular Objects, Abstract Data Types and Data Models: a Framework, Proc of the 10th VLDB Conf. (1984).

[Feld 86] Feldman P., Miller D., Entity Model Clustering: Structuring A Data Model by Abstraction, The Computer Journal, 29, 4 (1986).

[Kim 87] Kim W., Chou H.T., Banerjee J., Operations and implementation of Complex Objects, 3rd Int Conf on Data Engineering, Los Angeles (1987).

[Prie 87] Prieto-Diaz R., Classification of Reusable Modules, IEEE Software 4(1):6-16 (1987).

[Song 94] Song W., Capturing Semantics by Object Relativity, 6th International Conference, CAiSE'94: Advanced Information Systems Engineering, The Netherlands, Jun.(1994).

[Troy 91] de Troyer O., Schema Object Types: A New Approach to Modularization in Conceptual Modelling, Information Systems - Correctness and Reusability, Workshop IS-CORE (1991).

[Troy 93] de Troyer O., Janssen, On Modularity for Conceptual Data Models and The Consequences for Subtyping, Inheritance and Overriding. IEEE Proceeding Conf. Data Engineering (1993).

The OLYMPIOS Model for the Analysis, Specification and Design of an Information and Decision System

Ramzi GUETARI

LLP/CESALP BP. 806 - 74016 Annecy Cedex, France
Tel. +33 50 66 60 80 - Fax +33 50 66 60 20

Abstract

This paper presents a model, OLYMPIOS, for designing manufacturing firms' information and decision systems (IDS). This model proposes a designing method which ensures the continuity between different stages of the IDS life cycle. This method allows us to pass automatically from one stage to the following, so as to preserve the coherence obtained in the first stage. After the presentation of OLYMPIOS concepts, we present the different stages of an IDS life cycle according to our model, i.e. analysis, specification, design and implementation. We then present the automatic translation from one stage to another. Finally we illustrate the method with an industrial example.

Resume

Ce papier présente le modèle OLYMPIOS pour la conception de systèmes d'information et de décision (SID) d'une entreprise manufacturière. Ce modèle propose une méthode de conception qui respecte l'uniformité du modèle et assure la continuité dans les différentes étapes du cycle de vie d'un SID. Cette méthode assure un passage automatique d'une étape à une autre afin de garder la cohérence obtenue durant les premières étapes. Après une présentation des concepts du modèle, nous présentons les différentes étapes du cycle de vie du SID tel qu'il est vu dans OLYMPIOS, à savoir l'analyse, la spécification, la conception et l'implémentation. Nous présentons ensuite le passage automatique d'une étape à une autre. Le tout est illustré par un exemple tiré d'un cas industriel.

1 Introduction

The Object model has spread widely in programming languages over the last several years. The principles of this model have had a great influence on analysis and design techniques. However no existing method is able to manage the whole analysis-specification-design-implementation cycle, preserving the homogeneity of the model used in different stages and the coherence when passing from one stage to the following. We think that global management of the life cycle cannot be solved, with the existing state of knowledge, by one sole miraculous method, adaptable to every kind of application. We think moreover that the problem should be treated by a panel of methods dedicated to a particular domain.

For this reason we developed the OLYMPIOS model at the LLP-CESALP laboratory. This model covers the life cycle of every application in the field of "Information and Decision Systems for Manufacturing Firms". OLYMPIOS uses algebraic techniques, transformation rules and predefined entity organization to propose an original approach for the object oriented design of an information and decision system.

2 OLYMPIOS MODEL CONCEPTS

The information processed in an enterprise, which we call industrial information, is a complex datum. An information and decision system (IDS) must take this complexity into account. We propose to represent industrial information in four main facets :

- Data, describing the different entities handled by the IDS and the actions that they can perform or be subjected to ;
- Temporal properties of the different kinds of processes (including traceability of information) ;
- Organization, considered through information flows ;
- The economic facet, which describes the means of performance appraisal with relation to enterprise environment and objectives.

The OLYMPIOS model [Beauchêne 93][BHP 93][BHS 93] covers the different stages of such a system life cycle and proposes original solutions for its analysis, specification, design and realization. OLYMPIOS describes activities, taking into account the assigned objectives and the resource availability. The basic modeling elements are :

- An industrial information database, where products, resources, machines,... are described.
- Consumer-Supplier Information Systems (**CSIS**). A CSIS stands for an "atom" of organization. It is a generalization of the customer-supplier exchange relationship to every actor couple in the enterprise (men, machines, software). Every CSIS is associated with an objective, transforms resources and emits a satisfaction level.

- An Objectives Management System (**OMS**), whose role is to create a graph from expressed objectives, where every node is an objective associated with a CSIS.
- A Resources Management System (**RMS**), in charge of management and sharing of products and resources.
- An activation system (**AS**), producing action plans to organize processes, taking into account application, temporal constraints, and communications/synchronization between CSIS.

3 THE IDS LIFE CYCLE MANAGEMENT IN THE OLYMPIOS MODEL

The OLYMPIOS model covers the different stages of the IDS life-cycle. We use an algebraic approach for the four facets of industrial information so as to obtain a coherent (i.e. sufficiently complete and consistent) specification. The design stage enables us to conceive the information system from specification and by analyzing the "existing" system of the enterprise and its objectives. The result of this stage is a representation of the IDS using structured entities. The OLYMPIOS model introduces the uniformity of the model used from specification through to design. It uses some tools allowing the coherence of the system in the specification step and saves this coherence by automated translation from one stage to another.

3.1 Analysis and specification of the data facet

The data facet corresponds to the IDS functional and structural aspects and aims at representing the technical and technological data. We use Algebraic Specifications of Abstract data Types (ASAT) [Guttag 78] [Jacquenet 86] [Liskov 87] so as to have at our disposal efficient and simple proof techniques. An ASAT enables one to express an entity behavior in a high level formalism.

For a given entity, an ASAT is a triple $<\Omega,\Sigma,A>$ where:

Ω is a set of domains containing the domain of the entity values,

Σ is a set of operations on the entity,

A is a set of equations (axioms and preconditions) on these operations, which determines the entities' behavior and the relationships between them.

ASAT are automatically constructed from the entities' automata, which are the result of the analysis stage. This automatic construction is realized by some algorithms [Nkongo 90] developed in our laboratory and described in the final paper.

3.2 Specification of the "organization" facet

It starts from the analysis of the "existing system", which results (inter alia) in the identification of actors and their functions and objectives. Specifying organization consists in formally expressing identified objectives (in the "triple" form), and in constructing their associated CSIS from standard parametrized ASAT of organization [Beauchêne 93]. Simultaneously, one must elaborate the different graphs of objectives.

3.3 Specification of the "temporal" facet

The specification of the industrial information temporal facet uses a synchronous process algebra, directly derived from the SCCS calculus of R. Milner [Piard 93]. We specify four kinds of processes with this language :

- Chronological and event-based clocks, essential to specify synchronization and to measure temporal intervals ;
- Behavior of data facet entities, which are not completely determined by ASAT axioms ;
- Behavior of CSIS ;
- Activation plans, elaborated by the activation system from graphs of objectives and resources to schedule the CSIS.

3.4 Specification of the "economic" facet

This facet can not be specified independently of data and organization. Indeed it is shared between them, and the most important part is included in the organization facet. Work is still going on to sharpen the economic view of OLYMPIOS of the information system (with the help of performance indicators, fuzzy logic and project-based management approach).

3.5 Design stage

The OLYMPIOS model, in its design stage, is based on the class model. This model was extended in order to allow it to take into account all industrial information characteristics, notably real time ones. The result of the design stage is an organization of entities independent of the target programming language : OLSEN (**OL**ympios **S**tructured **EN**tity).

An OLSEN [Guetari 94] is composed of one part "class" and another part called "scenario" which indicates the interactions with its environment. The difference between an OLSEN and a classical object is the scenario which describes the temporal behavior generally absent in the standard class model. The OLSEN model is a "design object".

In this paper, we present only the specification and design of Activation System (AS part) and Resource Management System (RMS). The Objective Management System is the subject of a future publication.

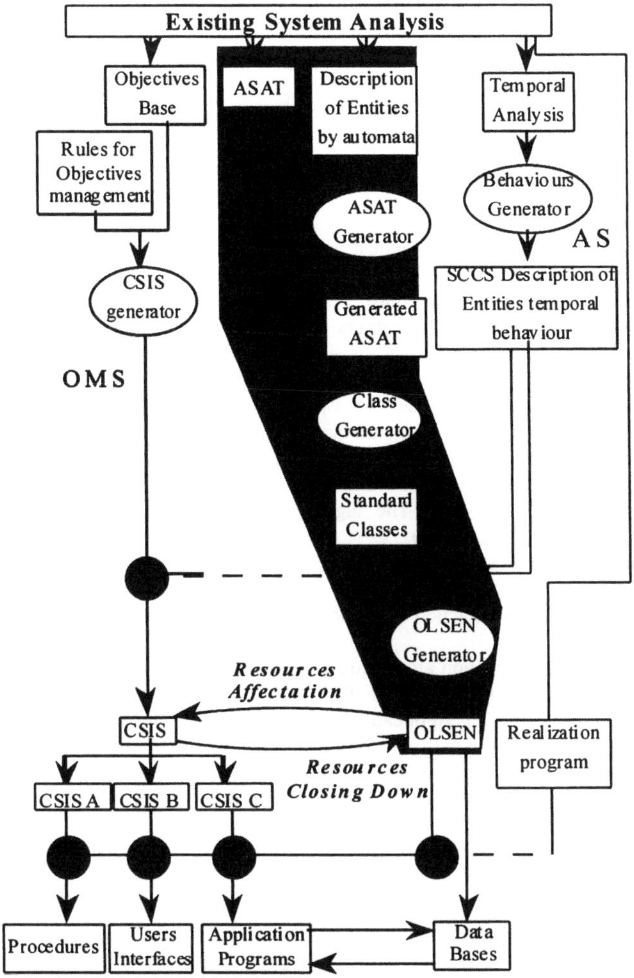

Figure 1. The Analysis - Specification - Design Cycle in the OLYMPIOS Model

CONCLUSION

In this paper we present the OLYMPIOS model and the three first stages of the IDS life-cycle : analysis, specification and design. We describe the specification by ASAT, and the tools which automatically generate ASAT from the analysis of data by automata. Then the OLSEN model is presented. We also display in detail the algorithms enabling one to pass from specification (ASAT and SCCS) to OLSEN and the automatic generation of OLSEN.

KEYWORDS

Object oriented design, industrial information, algebraic specifications, object structure, process calculus.

REFERENCES

[Beauch 93] Beauchêne D., L'information industrielle : définition et spécification. thèse de l'Université de Savoie, Dec. (1993).

[BHP 93] Beauchêne D., Haurat A., Piard F., Une méthode de spécification de l'information industrielle par types abstraits algébriques. Actes de ICO'93, Montreal Canada, 4-7 may (1993).

[BHS 93] Beauchêne D., Haurat A., Schweyer B., Designing an information system for a manufacturing enterprise under the aspect of a CIM approach : the model OLYMPIOS. Actes de APMS'93, 28-30 sept., Athenes, Greece (1993).

[Guetari 94] Guetari R., OLSEN : un modèle d'objets pour la conception de systèmes d'information industrielle. Rapport interne LLP/CESALP. Université de Savoie, Jan. (1994).

[Guttag 78] Guttag J.V., Horning J.J., The algebraic specification of Abstract Data Type, Acta Informatica, 10, 27-52 (1978).

[Jacqu. 86] Jacquenet J.P., Lescanne P., La réécriture", Techniques et Sciences Informatiques, 5 (6), 433-452 (1986).

[Liskov 87] Liskov B., Data Abstraction and Hierarchy. OOPSLA'87 Addendum to the Proceedings (1987).

[Nkongo 90] Nkongo T. , Spécification algébrique de types abstraits pour le modèle Olympios. Rapport de DEA Ingénierie Informatique de l'INSA de Lyon, Sept.(1990).

[Piard 93] Piard F., Braesch C., Application du calcul SCCS de Milner à la spécification de processus informationnels par types abstraits algébriques dans une entreprise manufacturière, Real Time Systems Conference, Paris (1993).

IMPEDIMENTS TO AN ELUSIVE DREAM : COMPUTER ACCESS TO NUMERIC DATA FOR ENGINEERING MATERIALS

Jack. H. WESTBROOK[1] *and J. G. KAUFMAN*[2]

[1] *Brookline Technologies, Ballston Spa , NY 12020 , USA.*
[2] *The Aluminum Association, 900 19th street, N.W., Washington D.C., 20006-2168, USA.*

ABSTRACT

Direct and easy-to-use access to a worldwide network of numeric material property databases has been an objective for the past fifteen years, and yet that result remains elusive. In this paper, the authors explore the technical and non-technical issues that have limited it, note some actions that may help meet the objective, and speculate on the nature of the next generation of materials data networks.

Among the key points brought out are (1) the relative complexity of numeric data on materials, (2) the problems in making an online system as readily accessible and easy to use as needed for end user searchers, and (3) the economic squeeze between the high costs to achieve success (points 1-2) and the requirement to become self-sufficient and generate a financially self sufforting system. A significant added problem is the lack of consistently high quality data and databases or networks thereof, sufficiently comprehensive to meet the needs of a broad range of potential users with varying needs.

RESUME

L'accès direct et facile à un réseau mondial de banques de données sur les propriétés des matériaux constitue un objectif vieux de quinze années et cependant evasif. Dans cet article les auteurs explorent les objectifs techniques et non techniques sources de limitations, évoquent quelques actions susceptibles d'être utilisées pour avancer et spéculent sur la nature d'un prochain réseau sur les données des matériaux.

Parmi les points clés évoqués , signalons (1) la complexité relative des banques de données numériques de matériaux (2) le problème de la simplification des systèmes en ligne conduisant à une utilisation finale aisée (3) l'impasse financière liée à la recherche des coûts élevés(points 1-2) et la recherche d'un système autofinancé. Un autre problème important est celui de l'insuffisance de données de haute qualité et des bases de données ou réseaux associés pour faire face à la diversité des besoins potentiels des utilisateurs.

1 INTRODUCTION

Computer access to a broad distributed network of numeric material property data for engineering has been envisioned as a great desideratum of the engineering profession for at least thirty years and yet, despite major efforts in that direction, it has yet to be realized in a comprehensive, permanent and self-supporting fashion. That is in marked contrast to the evident market success of on-line bibliographic reference sources in various fields, most notably the legal and chemical fields, and also of online numeric data in the financial field. It is the purpose of this paper to aid in diagnosing the reasons for these differences, and especially the reasons for the limited success in networks of numeric engineering data, with the motivation of aiding in identifying pointers for future activity in this area.

2 THE EXPERIENCE, BRIEFLY

Over the past five years, a number of previously very active programs aimed at networking numeric mechanical, physical, and chemical data for materials of many types, including pure substances, compounds, mixtures and engineering structural materials, have ceased to move forward. Two of the most significant of these were actually abandoned in their original concept. The Commission of European Communities, after several years of exploring the requirements and establishing the base systems (1), abandoned hope of actually building a distributed network of numeric databases even on a demonstration basis.

In a more nearly commercially successful effort, the National Materials Property Data Network, Inc., together with an associate, the American Chemical Society through its Chemical Abstracts Service organization, created the Materials Property Data Network (MPD Network) and the companion Chemical Property Data Network (CPD Network); both were made available on STN International for about two years (2). They are no longer available, primarily because costs exceeded revenues over too protracted a period without a fully satisfactory quality of service being available. Future implementation of these concepts is still a matter of study by the original sponsors of the MPD Network approach.

It is useful to explore some of the major reasons for the limited success of these endeavors, noting those aspects of success and more importantly the factors that precluded successful permanent and self-supporting operations.

3 THE TECHNICAL ISSUES

Admittedly, most of the more successful online network services have been substantially easier to accomplish technically, and therefore less costly to establish and maintain, than the network of material property databases. Once the intellectual task of creating references or abstracts is completed, providing electronic access to textual reference data involves relatively simple loading processes and also straightforward string search techniques. Numeric engineering materials data, on the other hand, even without the complexity of networking, involve several additional levels of complexity as discussed in two prior CODATA Workshops (3,4) and a series of ASTM symposia proceedings (5-7).

- Numeric databases for materials property data must be created *de novo*; even conversion of printed handbook data to electronic form is a very complex task (8-10)

- Each numeric engineering data element may contain a variety of complex units each of which must be linked with the appropriate property value at all times; without the units, the numeric values themselves are worthless.

- Numeric engineering data frequently mix greatly differing orders of magnitude within a single data unit. They may readily range from 10**6 to 10**-6.

- Numeric engineering data must regularly be searched as numeric values or ranges of values, so that rather sophisticated loading and search software is required to create, maintain and utilize the databases.

- Many of the numeric values may be directly dependent upon one or several independent variables such as time, temperature and pressure. As with units, the values of all associated variables, themselves having units, must be retained with the property itself, or its interpretation and utility will be greatly diminished.

- Single property number lookup is a relatively simple and unsophisticated use of a materials database. More complex uses such as material selection, process control, or component design demand features that add to the net value of and hence incentive to use databases, but they also add to the complexity and cost of the database development.

The effect of these complexities can readily be an order of magnitude greater cost to build, maintain, and update these databases, above and beyond the substantial initial investment in the search software to perform the sophisticated queries required. The upfront investment in such systems is substantial, and the revenue that must be obtained in their use is substantial if they are to be self-supporting. User aspects, as discussed by Kaufman (11), Rumble (12), and Hossain and Barry (13), present some additional issues with a technical component: access to online data systems and the user front end to the system. With regard to access, it is appropriate to note that until very recently (and still today for many engineers and material scientists), gaining online connections via computer modem telephone communication software, and the diversity of telephone access routes has made achieving a good online connection a daunting process. Even when user-oriented software, like the MPD Connection, was made available, users often found a number of roadblocks to smooth online access.

The result was frustration and an unwillingness to try again in many cases. Once online, the problems are not over because easy-to-follow directories to materials databases have not been generally available except in specialized systems such as MPD Network.

Another problem arises because numeric material property databases are best searched by the engineers themselves, as contrasted to reference librarians or professional searchers. Therefore there is a need for "user friendly" search software; often a menu-driven relatively intuitive front-end has been the objective of the creators of such systems. This also adds to the investment needed for such a system. Even when a considerable investment is made, we find that what is intuitive to one searcher or to a searcher with one mission in mind is not intuitive to another who may have a completely different mission in mind. The user interface contains another hazard. Lack of "user friendliness" in a bibliographic database may only pose an inconvenience; but, in contrast, in an numeric materials database the consequence may be a disastrous technical decision or a legal liability. Despite a substantial investment in the user interface, including the capability to ask sophisticated questions, the MPD Network interface never reached the facile use desired, especially for the occasional searcher, and that is a fair characterization of most engineers and material scientists.

In either a distributed or centrally hosted network of such databases, there is the added problem of variations in data structure and terminology, as described by McCarthy (14) and Westbrook and Grattidge (15). In order to deal with this, a thesaurus is a prime need for making the databases work well together. In the distributed system, such problems are amplified by the number of interfaces that must be dealt with transparently. The MPD Thesaurus (2,11) was a direct response to this need and was well developed, but added great intellectual and programming effort for maintenance and updating. This matter is discussed further under data issues. The net effect of these various levels of complexity, generally underestimated in importance by the creators of such systems, is the extremely high cost of an easily accessible and broadly useful network of databases on the properties of engineering materials.

4 The Non-technical Issues, Including Economics

There are several non-technical issues, reviewed in part by Rumble (16) and Barrett (17) that have limited the success of materials data networks to date, including:

- Lack of commitment by government agencies - Despite the benefit of rather logical ties to the organizations building the networks, little institutional support was made available. Government agencies, anxious to use their funds for long range, high impact research, did not buy into the fact that computerizing and networking existing data would be a valuable contribution of broad utility and high leverage. It was easy to conclude that private industry should handle this seemingly simple task. Only the Standard Reference Data Program at the National Institute of Standards and Technology (NIST) saw the opportunity and tried to add support, but was hampered by tight budgets and lack of cooperation from other agency groups.

• Lack of interest by technical societies - The logical technical societies that might lead the materials database network effort interpreted the MPD Network approach not as an opportunity but a threat to their substantial revenues from print information products. Even the fine support provided by the one group, the American Chemical Society, within the infrastructure of their own network, was not sufficient to sustain the development effort alone without more cooperation from other materials groups.

• Lack of staying power by private industry - More than sixty (60) representatives of private industry provided some level of financial support for the MPD Network, but most of those sixty provided $1,000 to $5,000, a few provided up to $25,000; and only one industry group provided $100,000 or more. Given the projected complexity and breadth of the desired service, verified by seven industrial workshops cosponsored by MPD Network and SRD/NIST, most of the contributions were token support, sometimes overtly stated as charitable in nature.

• Perception that materials data should be "free" - The engineering and material science communities were philosophically unprepared to pay significant fees for materials data, even with the attractive concept of simultaneously searching all of the world's materials data with one query. The perception is that information in books is free, and the added ability to manipulate data and initiate complex material queries via electronic access to computerized databases is typically not equated with value added.

• Data "in-hand" vs. data "online" - There was a continuing desire to have the databases and search systems in hand, not in the intangible, seemingly uncontrollable cyberspace of the online world. Related to that is the fear of running up huge bills with mistaken or worthless searches in online systems controlled by others. Even the introduction of subscription pricing, which provided budgetary control of costs, did not satisfy this concern so we believe that control of the data is a more important factor than the cost concern.

• Costs - The high costs of building effective systems (computerization of data, incorporation of ancillary data and descriptors, graphical presentation, database management, and other software, etc.) were not fully recognized; hence efforts were consistently undercapitalized.

5 Content Related Issues

In addition to the technical and non-technical economic issues noted above, there is one additional class of problem or limitation that we choose to isolate, as it may be the most difficult to deal with and the "fatal flaw" in the current schemes - that of the limitations of the available data. History suggests that had the data been sufficiently useful, the limitations of cost and ease of access would have been dealt with by the users.

• Comprehensiveness - While certain classes of materials and certain specific properties are relatively available in computerized materials databases that could be and were in fact networked (a total of twenty five databases were included in the MPD and CPD Network systems) there is not yet the consistently broad coverage of materials and properties across the entire matrix that might be desired for many applications.

• Standardization - Even where reasonable completeness is provided by the combination of several databases, they tend not to be in parallel structures, not to use standardized terminologies and definitions, and not to have sufficiently cohesive units systems such that they can be easily searched jointly even with the help of inter-database thesauri and units conversion (18-22).

• Quality and reliability - Because most databases are built with specific, perhaps widely varying needs and applications in mind, it is not uncommon for different databases to have been developed with different data reliability standards in mind. Some have statistically reliable design properties; some have typical or average properties; others have individual test results. They really are not comparable and should not be searched together. If they are included in a single integrated system, the differences in data reliability need to be made very clear, preferably by the inclusion of a data quality index right in the individual databases. Locating a large group of databases all having the same level of reliability has not yet proven a practical reality. These issues have been further discussed by Barrett (23) and Kaufman (24).

Realistically , databases that optimize all of these factors for all of the materials and properties that might be desired simply are not available for networking. Even if proprietary databases could be included (and that in itself is a limitation, as they typically can not be made available) complete coverage is unlikely.

6 Conclusion : Where Do We Go From Here?

It is apparent from the experience gained, especially from about 1985 to the present, that there are no simple solutions readily available that will satisfy the desire for a single network, easily accessible and readily usable, of all of the world's numeric materials databases for engineers and materials scientists.

Recent surveys of user needs, both through market studies by MPD Network in cooperation with the American Chemical Society and by the authors in preparing this paper, show a broad use of computers and considerable need for electronic access to data, but very limited awareness and use of computerized online databases. This may change with the rapid growth and increasing capability of INTERNET and associated access services, but all of the technical and economic issues noted above will remain.

There are some things that would help bring reality closer to the objective of worldwide network access to high quality materials data:

- More aggressive standardization
- Improved (more intuitive to the user) access and search software
- More rapid system response to queries
- Better links to application software

If these issues can be dealt with, and if a suitable sponsor for the activity can be identified, it seems likely that many of the other issues such as the approach to marketing will be dealt with.

Recent changes in technology have increased the likelihood that any future incarnation of the materials database network concept is more likely to involve CD-ROM systems or workstation systems than online distribution. CD-ROM was very seriously considered for the MPD Network in 1985, but at that time compact disk systems had not approached a practical level of acceptance and standardization that made reliance on that approach feasible. It may be the most logical approach for the next generation.

Keywords

Network , database , numeric , materials , properties , thesaurus , interface

References

[1] Swindells N., Waterman N., Kröckel H. (eds), Materials Information for the European Communities, CEC, Luxembourg (1990).

[2] Kaufman J. G., Online Access to Worldwide Sources of Materials Performance Data - 1991 Update on the Materials Property Data Network (MPD NetNork), in Computerization and Networking of Materials Databases, ASTM STP 1140, Reynard and Barry (eds), ASTM, Philadelphia (1992).

[3] Westbrook J. H., Rumble J.R. Jr. (eds), Computerized Materials Data Systems, Proceedings of a Workshop, Faifield Glade, TN, November 7-11, 1982, National Bureau of Standards, Washington, DC (1983).

[4] Westbrook J. H., Behrens H., Dathe G., Iwata S. (eds), Materials Data Systems for Engineering, in Proceedings of a CODATA Workshop held at Schluchsee, Germany , September 22-27, 1985, FIZ, Karlsruhe, FRG (1986).

[5] Glazman J. S., Rumble J. R. Jr. (eds)., Computerization and Networking of Materials Databases, ASM STP 1017, ASTM, Philadelphia, PA (1989).

[6] Kaufman J. G., Glazman J. S. (eds), Computerization and Networking of Materials Databases, Vol 2, ASTM STP 1106, ASTM, Philadelphia ,PA (1990).

[7] Barry T. I., Reynard K. W (eds), Computerization and Networking of Materials Databases, Vol. 3, ASTM STP 1140, ASTM, Philadelphia, PA (1992).

[8] Grattidge W., Capture of Published Materials Data, in Computerization and Networking of Materials Databases, 151-174, ASTM STP 1017, Glazman and Rumble (eds), ASTM, Philadelphia, PA (1989).

[9] Westbrook J. H., Grattidge W., The Role of Metadata in the Design and Operation of a Materials Database, in Computerization and Networking of Materials Databases, Vol 2, 84-102, ASTM STP 1107. ASTM, Philadelphia. PA (1991).

[10] Grattidge W., Lund W. B., Westbrook. J. H., Problems of Interpretation and Representation in the Cornputerization of a Printed Reference Work on Materials Data, in Computerization and Networking of Materials Databases, Vol 3, ASTM STP 1140, Barry and Reynard (eds), ASTM, Philadelphia, PA (1992).

[11] Kaufman J.G., Increasing Data System Responsiveness to End-User Expectations, in Computerization and Networking of Materials Databases, Vol 2, 103-113, ASTM STP 1106, Kaufman and Glazman (eds), ASTM, Philadelphia (1991).

[12] Rumble J. R. Jr., Access Paths for Materials Databases, Computerization and Networking of Materials Databases, in Vol 2, 132-141, ASTM STP 1106, Kaufman and Glazman (eds)., ASTM, Philadelphia, PA (1991).

13] Hossain M.K., Barry T.l., The Needs of Users and Their Response to Materials Data Banks, in Computerization and Networking of Materials Databases, Vol 3, ASTM STP 1140, Barry and Reynard. (eds), ASTM, Philadelphia, PA (1992).

[14] McCarthy J.L., The Automated Data Thesaurus: A New Tool for Scientific Inforrnation, in Scientific and Technical Data in a New Era, Proc. of the 11th CODATA Conference, 260-264, Glaeser (ed), Hemisphere Publishing Corp., New York (1990).

[15] Westbrook J.H., GrattidgeW., Terminology Standards for Materials Databases, in Computerization and Networking of Materials Databases, Vol 3, ASTM STP 1140, Barry and Reynard (eds), ASTM, Philadelphia, PA (1992).

[16] Rumble J.R. Jr., Socioeconomic Barriers in Computerizing Materials Databases, 216-226, ASTM STP 1017, Glazman and Rumble (eds), ASTM, Philadelphia, PA (1989).

[17] Barrett A.J., The Benefits and Economic Consequences of Materials Property Databases, in Vol 2, , Computerization and Networking of Materials Databases, 17-25, ASTM STP 1106, Kaufman and Glazman (eds), ASTM, Philadelphia, PA (1991).

[18] Kröckel H., Reynard K.W., Rumble J.R. Jr. (eds), Factual Materials Data Banks - The Need for Standards, VAMAS Technical Report No. 2, VAMAS, July (1987).

[19] Kaufman. J.G., Standards for Computerized Material Property Data -ASTM Committee E49, in ASTM STP 1017, 7-22, Glazman and Rumble (eds), ASTM, Philadelphia, PA (1989).

[20] Rumble J.R.Jr., Standards for Materials Databases: ASTM Committee E 49, in Computerization and Networking of Materials Databases, Vol 2, 73-83, ASTM STP 1106, Kaufman and Glazman (eds), ASTM, Philadelphia, PA (1991).

[21] Swindells N., Kröckel H., Standards for Materials Databases: An International Concern, in Data for Discovery, Proceedings of the 12th International CODATA Conference, 276-285, Glaeser (ed), Begel House, New York (1992).

[22] Reynard Keith W., Standards for Materials Databases - National and International Programmes. Do They Provide for Data Users' Needs?, in Computerisation and Networking of Materials Databases, Vol 3, ASTM STP 1140, Barry and Reynard (eds), ASTM, Philadelphia, PA (1992).

[23] Barrett Anthony J., Social Implications of Data Quality, in Scientific and Technical Data in a New Era, Proceedings of the 11th CODATA Conference, p339, Glaeser (ed), Hemisphere Publishing Corp., New York (1990).

[24] Kaufman J. G., Quality and Reliability in Materials Databases, in Computerization and Networking of Materials Databases, Vol 3, ASTM STP 1140, Barry and Reynard (eds), ASTM, Philadelphia, PA (1992).

SCIENTIFIC DATABASES: TENTATIVE ANALYSIS OF PRODUCTIVITY

Jean Loup DELCROIX

Université de Paris-Sud, Centre d'Orsay Laboratoire de Physique des Gaz et des Plasmas, Centre de Données GAPHYOR, Paris, France

ABSTRACT

The information in a given Scientific Database can be divided into three categories: factual, numerical and bibliographic. It is then possible to define two ratios : a global one, Information/Cost called the Productivity, and a bibliographic one, Integrity/Volume called the Efficiency. This analysis can be used to evaluate some possible improvements or simplifications of a given Database. The basic data for the calculation of Efficiency are some statistics about the authors' names. This ratio can then be optimized by limiting the number of registered authors in a given paper and moreover truncating their names. This exhibits an interesting dilemma between the standard methods of documentalists and much simpler descriptions that are quite satisfactory for Database users.

RESUME

L'information contenue dans une Base de Données Scientifiques peut être séparée en trois catégories : factuelle, numérique, et bibliographique. On peut alors définir et évaluer deux rapports : un rapport global Information/Coût baptisé Productivité, et un rapport bibliographique Intégrité/Volume baptisé Efficacité. Cette analyse peut servir à évaluer le bien-fondé d'améliorations ou de simplifications éventuelles d'une Base de Données. Les données de base pour le calcul de l'Efficacité sont des statistiques sur les noms d'auteurs. Le rapport peut alors être optimisé en limitant le nombre des auteurs enregistré pour une référence donnée, et aussi en tronquant leurs noms. Cela fait apparaître une contradiction entre les méthodes standard des documentalistes et des descriptions beaucoup plus simples tout à fait satisfaisantes dans une banque de données.

1 INTRODUCTION

Producers of scientific databases must now pay more attention to their costs of production and try to minimize them. In the present situation the main cost is manpower, and essentially the work of science experts who are in charge of collection and evaluation of data. One must also mention that these people do not find this activity very gratifying, and consequently there is a strong need to simplify their work. We are then now at a phase of this business where one must try to increase the productivity of database compilers, without decreasing the quality of the product from the point of view of users. We describe such an attempt in the present paper. This will be done by trying to evaluate and then to optimize an Information/Cost ratio that we call the Productivity. To be sure this is only a semi-quantitative approach. In its present state it provides nevertheless some guidelines to decide whether some possible improvements or simplifications of a given database are sensible or not.

In the second part of this paper we perform a statistical study of authors' names and of the possible simplifications of this part of a bibliographic reference. We define a ratio called the Efficiency which is useful to evaluate possible simplifications of the bibliographic part of the information. The limits and future development of this work are discussed in the Conclusion.

2 STRUCTURE OF GAPHYOR DATA

2.1 Factual, numerical, and bibliographic information

A new structure of data in GAPHYOR was adopted during 1993-1994 in conjunction with a new implementation on an autonomous Unix work-station. It is described in the new *Manuel Interne GAPHYOR* [1]. In this structure the data are specified by a list of 49 descriptors forming four groups :

- *Factual data* (33 descriptors). Recalling that GAPHYOR covers Physics and Chemistry of atoms and small molecules in gases, the factual data describe the three possible reactants in an initial state, the three possible reactants in the final state, the category of process, and specific information on the nature of the data.
- *Numerical data* (3 descriptors). These includes the energy of the reactants, the values of the characteristic quantities in the process (energy levels, cross sections, reaction rates, transition probabilities, ...), and some additional information if necessary.
- *Bibliographic data* (15 descriptors). These include the standard bibliographic references including up to ten authors.
- *Internal data* (3 descriptors). These are not for the database users (registration rm number, ...)

2.2 Statistical analysis of a specimen of GAPHYOR

We shall base the discussion in the present paper on a specimen called "GAPHYOR-94" including 7600 entries taken from four sources :

- PR/A : Physical Review A
- ZP/D : Zeitschrift fur Physik D (Atoms, Molecules, Clusters)
- JRD : Journal of Physical and Chemical Reference Data
- DIV : Books, Reports and Conferences proceedings.

3 EVALUATION OF INFORMATION IN THE DATABASE.

The quantity of useful information in the database can be evaluated in two steps. The detailed calculations are summarized in reference [2]. We describe below the methods and the main results.

3.1 Purely quantitative evaluation

We consider the raw information in every group of data (factual, numerical, bibliographic, internal) as just the number of included bytes (or characters). This sensible evaluation will be improved in the next subsection. Let us then call *n(fact), n(num), n(bib), n(internal)* the average length in bytes of the four groups of data in an entry. The mean values of these quantities for GAPHYOR-94 are shown in Table 1.

	PR/A	ZP/D	JRD	DIV		Σ = (GAP-94)	
n(fact)	17	17	27	21		22	**29%**
n(num)	8	3	20	6		12	**15%**
n(bib)	50	51	28	55		43	**56%**
–	76	72	76	82		76	100%
n(internal)	10	10	10	11		10	
—	86	83	86	93		87	

Table 1. GAPHYOR specimen ("GAPHYOR-94") statistical properties.

3.2 Value corrected Information

It is clear that all the entries in a database have not the same value for users. It is then necessary to correct the preceding evaluation by some factors taking into account this value. We shall then calculate this corrected information by the following formulae :

$$In(fact)=n(fact)*V(fact)*V(bib) \quad (1)$$

$$In(num)=n(num)*V(num)*V(bib) \quad (2)$$

$$In(bib)=17.6*V(bib) \quad (3)$$

where n is, as before, a length in bytes, V a value index and In a corrected information quantity. Note that in the last of these formulae we have set $n(num)=17.6$: this means that we consider that the useful bibliographic information includes only the name of the first author of average length 7 (cf. Section 5).

The calculation of the value indexes is somewhat arbitrary, but with a good knowledge of the database field, and some discussions with the users, sensible choices can be made. In the case of GAPHYOR these have been discussed in the GAPHYOR HANDBOOK [3]. For a full description of all these indexes, we send the reader to references [2] and [3]. Let us just say that the indexes are chosen so as to take account of the following qualities of the paper and the entry :

- absolute values of data in the paper
- complete description of the reaction scheme in the entry
- low (thermal) energies of reaction
- experimental results
- interesting processes
- numerical values of energies in the entry
- numerical values of data in the entry
- additional remarks on the data in the entry
- recent publication
- first class journal.

The formulae above were then used to evaluate the average value-corrected information in an entry of GAPHYOR-94. The average results per entry are :

In(fact)/entry=9 520 GCU

In(num)/entry=7 270 GCU

In(bib)/entry=305 GCU

In(tot)/entry=17 095 GCU

We get then for the total useful Information in GAPHYOR-94 :

In(tot)=1.23 10^8 GCU (4)

Note that the units used(GCU : GAPHYOR Count Units) are arbitrary, but that the choice of value indexes is such that the two groups of factual and numerical information have about the same weight, while the bibliographic information is more than twenty times smaller.

4 Evaluation of Cost of Production

The "Cost of the Database" is the sum of fixed costs, essentially C(inv) (investments) and C(sal) (fixed salaries) and variable costs that we shall call the "Cost of Production". To evaluate this last one, we start from the statistics of Table 1, and of some economic parameters such as the salary per hour of science experts, the number of bytes entered per hour of work, the cost c of entry per byte, the cost of processing per byte, and the cost of documentation sources per page.

Note that in this paper we have neglected the costs of processing and documentation because we want to concentrate on the cost of work by science experts in charge of collection and evaluation of data (in any case the cost of processing is less). The following scheme is then based on the three formulae :

$$C(vol)=C(vol)\_0+sum\ C(ref) \qquad (5)$$

$$C(ref)=C(bib)+C(ref)\_0+sum\ C(fic) \qquad (6)$$

$$C(fic)=[n(fact)+n(num)]c \qquad (7)$$

Formula (5) expresses the cost *C(vol)* of extraction from a volume as the sum of the cost *C(vol)_0* of scanning for relevant references, and the cost *sum C(ref)* of extraction from these relevant references. Formula (6) similarly expresses the cost *C(ref)* of extraction from a reference as the sum of the cost *C(bib)* for typing the bibliographic information, the cost *C(ref)_0* of scanning for relevant data, and the cost *sum C(fic)* of typing these data. Finally Formula (7) expresses the cost of producing a factual and numerical entry.

The detailed calculations of "Cost of Production" are given in [2]. They include, in the evaluation of the cost of typing the bibliographic information, reduction factors to realize that this information is typed only once per reference or even once per computer screen. The final result for the cost of production of GAPHYOR-94 is :

$$C(tot)=48{,}682\ F$$

Combining this with (4) we obtain the ratio Information/Cost that we shall abbreviate the Productivity P as :

$$P=In(tot)/\,C(tot)=2{,}670\ GCU/F$$

Note that the accuracy of all the figures above is totally fictitious. We keep these figures just for later comparisons between various models : the differences or ratios between the results are indeed much more significant than the absolute values.

5 Applications of Productivity Analysis

5.1 Global analysis of two models

One can use the "Productivity Analysis" to decide whether some improvements or simplification of a given database are sensible or not. As simple examples we shall discuss first two models of modification :

- Model (N0) where we assume that all the *numerical information is omitted.* The qualitative result is obvious : there will be a significant decrease of Productivity, since the numerical information is very useful.
- Model (0,0) where we assume that in the bibliographic information the *authors* and laboratories are omitted. In this case we expect an increase of the productivity, since the work of experts is alleviated, while the loss of Information is very small, since the authors' names are somewhat redundant with the Journal, Volume and Page information.

The productivity analysis of these models has been made in the case of GAPHYOR-94. The comparison with the original database is shown in Table 2. We note that this productivity analysis confirms the predictions we made. Moreover one sees that the increase of Productivity in Model (0,0) (omitting authors) is very small indeed. But one must note that :

- The preceding analysis is global, in the sense that it includes all the Journals of GAPHYOR-94. This is not quite satisfactory because a given expert scans usually a few specific Journals and the conclusions of the productivity analysis may vary from one Journal to another. To improve it we analyze below the Productivity for each separate source .
- The problem is not strictly one of cost. It is also psychological, because science experts feel that copying up to ten authors' names is a rather stupid kind of work. This shows then the limits of Productivity analysis : this is why we shall develop in the next section an analysis of the Efficiency in the registration of bibliographic information.

Model	GAP-94	N0	(0,0)
Relative Productivity	**1.00**	**0.44**	**1.05**
Productivity Variation	**0%**	**-56%**	**5%**

Table 2. Comparison of total GAPHYOR 94, Models N0 and (0,0). (Relative Productivities are obtained by comparison to GAPHYOR-94).

5.2 Separate journals analysis

Let us now improve our Productivity analysis by applying it separately to the three Journals scanned in GAPHYOR-94 . The results are shown in Table 3 which includes also for every source the Model (0,0) transformation (no authors' names).

One notes that for PR/A and ZP/D the increase in productivity obtained by omitting names is significant, while it is substantially zero for JRD, where there are many entries per reference.

One could also use this kind of analysis to decide whether it is worth while to enter into the database some rather inefficient Journals. For GAPHYOR we want to make this analysis in the near future, but for this we must get a safer evaluation of the Value indexes of all the journals.

	PR/A		ZP/D		JRD	
	GAP-94	(0,0)	GAP-94	(0,0)	GAP-94	(0,0)
Relative Productivity	**0.46**	**0.57**	**0.47**	**0.54**	**2.19**	**2.19**
Productivity variation	**0%**	**23%**	**0%**	**16%**	**0%**	**0%**

Table 3. Comparison of GAPHYOR 94 and Model (0,0) for separate journals. (Relative productivities by comparison to total GAPHYOR-94).

6 STATISTICS ON AUTHORS NAMES

Let us first state the problem we are going to study : we have seen that in a typical database the raw volume of the bibliographic Information may represent a large fraction of the total raw volume. In this part of information we must however distinguish first the year, journal, volume, page information which is essential to the users : we can call it the Localization (or retrieval) Information (LINF).

In addition, bibliographic information includes authors' names, laboratories of these authors : we can call it the Personalization Information (PINF) ; this second part is often responsible for a rather useless inflation of the total raw bibliographic stored information. To get a better feeling about this issue we have performed in GAPHYOR-94 a statistical analysis on authors names. The results are shown below.

6.1 Number of authors signing a paper jointly.

Most scientific papers in journals are now written not by a single author, but by a group of authors signing jointly. There are even some pathological cases where the group may include up to a hundred people or more, especially in fields involving work with large equipment (high energy particles, plasmas, ...). The statistics on numbers of authors per paper is shown in Figure 1.We note that about 13% of papers are signed by more than five authors. One can also evaluate the average number n(AUT) of authors per paper, and find : $n(AUT) \cong 3.42$

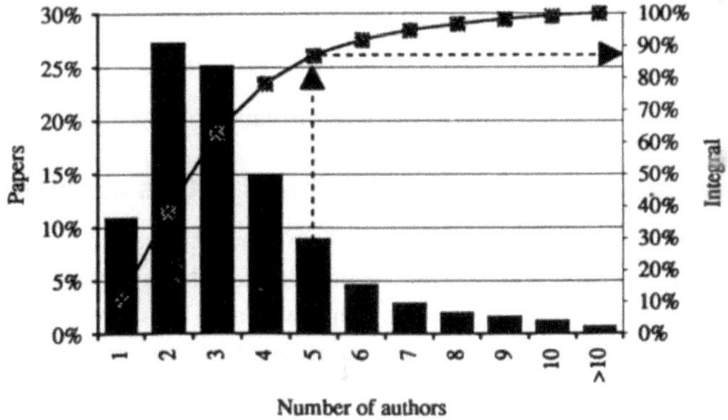

Figure 1. Number of authors signing a paper jointly.

6.2 Length of authors names

The standard use in bibliography is to cite the authors' names followed by initials of their first names. We followed it in GAPHYOR-94 writing for example "Dupont J L". The resulting statistics of the lengths of these names is shown in Figure 2. We note that there are about 13% of names with more than twelve characters.

One could possibly decide to omit the initials : the corresponding statistics are also shown in Figure 3(b). One notes then that there are about 12% of names without initials and with more than nine characters. One can also evaluate the average length *L(AUT)*, and one finds :

$$L(AUT) \cong 9.70 \text{ with initials}$$

$$L(AUT) \cong 7.03 \text{ without initials}$$

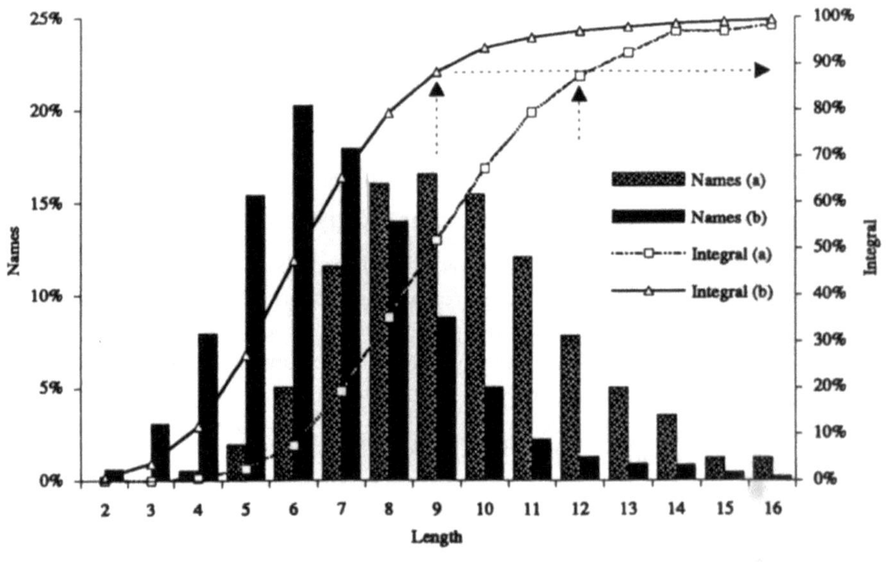

Figure 2. Length of authors' names, (a) with initials, (b) without initials

6.3 Homonyms created by suppression of initials or truncation.

Omission of initials creates homonyms : one finds that this *omission of initials* creates about *10% of homonyms*.. One could also decide to truncate names (including initials) to a fixed number of characters. This will also generate artificial homonyms. The shorter the truncation the larger the number of homonyms. The statistics of this homonym creation is shown in Figure 3. We note that truncation to five characters creates about 10% of homonyms.

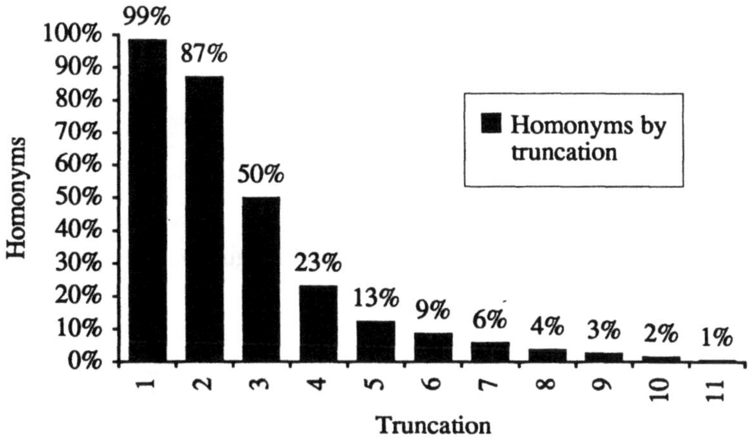

Figure 3. Creation of Homonyms by Truncation

7 SIMPLIFICATION OF BIBLIOGRAPHIC REFERENCES

Using the statistical results above we can now study possible simplifications of bibliographic references. As we noted before, the PINF Information is not very useful to users of the database, since it is somewhat redundant with the LINF part. An extreme simplification of the Bibliographic Information would be just to omit PINF : this is the model (0,0) that we discussed in preceding sections.

To make less drastic simplifications we can decide that they are defined by the two following rules :

- Omission of initials : this is standard use for scientists who speak for instance of a paper of "Dupont" published in 92.
- Limitation to n_A of the number of authors kept in the Database for every entry.
- Truncation to L_A characters of authors' names.

The result of these rules is to alter somewhat the PINF Information (note again that this is not a big nuisance). This can be specified by defining the Integrity ratio :

I= number of totally preserved references/total number of references

As an example Figure 4 shows the variation of I with the truncation length L_A, and for n_A= 3 authors. We note that it increases rapidly from zero to 50% obtained for L_A=8, and then more slowly to an asymptotic limit 0.63 corresponding to no truncation at all.

Since this perfection is rather costly we shall now introduce an Integrity/Volume ratio: $e=I/(n\_A*L\_A)$

that we call the Efficiency. We shall see that this ratio is maximum for :

$$n\_A=3 \qquad (8)$$

$$L\_A=8 \qquad (9)$$

We introduce then the "relative" efficiency :

$$E=e(n\_A,L\_A)/e(3,8)$$

The values of this relative efficiency are shown in Figure 5.

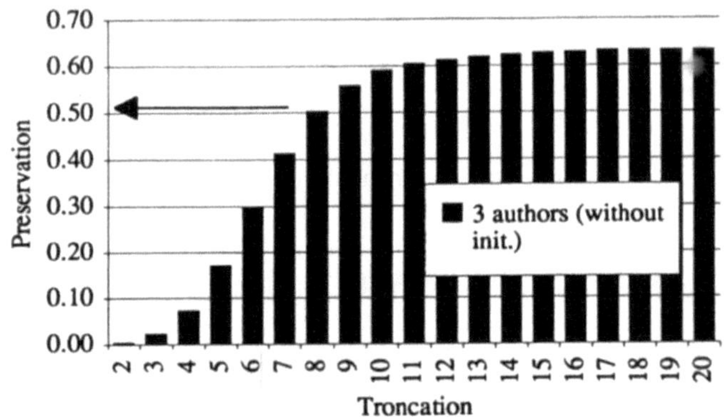

Figure 4. Integrity ratio for n_A= 3 authors as a function of the truncation length L_A

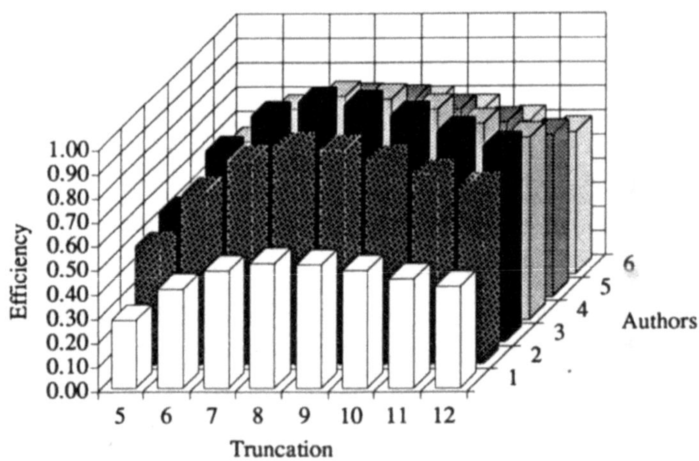

Figure 5. Efficiency (relative) as a function of n_A and L_A (Note the maximum for n_A=3, L_A=8).

These results show that a sound choice to simplify Bibliographic Information is given in formulae (8) and (9). This is a big simplification for the experts. But we must note that the improvement is more psychological than economical, since the increase in Productivity is rather small as shown in Table 3.

	PR/A		ZP/D		JRD	
	94	Model (3,8)	94	Model (3,8)	94	Model (3,8)
Relative Productivity	**0.46**	**0.52**	**0.47**	**0.51**	**2.19**	**2.19**
Productivity variation		**13%**		**8%**		**0%**

Table 3. Comparison of PR/A, ZP/D, JRD, in GAPHYOR 94 and in Model (3,8). (Relative Productivities are obtained by comparison to total GAPHYOR 94).

CONCLUSION

- The separation in a given Database of Information into three categories : factual, numerical and bibliographic and the methods of evaluation of the amount of useful information and of production cost developed in this paper provide some interesting guidelines for an optimum design of a given Database. They are at present not accurate enough. In the future we shall refine the evaluation of value indexes, especially of journals. The study shows also that the problem of simplifying the work of experts in charge of data collection and evaluation is motivated by psychological rather than economical reasons.

- Some statistics on authors' names have been performed. In spite of the limited volume of the specimen used, these are of quite general interest. In particular they provide guidelines to simplify the bibliographic part of the information by limiting the number of registered authors in a given entry, and moreover truncating their names.

- The proposal of limiting the number of registered authors, and truncating their names is in opposition with documentation technology usage. It just means that the objectives of scientific database producers and librarians are different : the first try to produce rather simple systems, permitting an easy and complete retrieval of the scientific information, while the latter try to produce truly exhaustive systems of documentation. That is life !

- The contradictions above are somewhat resolved when in database design one distinguishes between stored information and diplayed information. At the cost of slightly more work by the experts, the stored information can follow the strict rules of the art. But in any case the displayed information should follow the rules of productivity and efficiency that we discussed in Sections 5 and 6 above. This is probably the main interest of the present study.

- Since we finally find that the main interest of this study is to provide guidelines for the simplification of *displayed information*, we shall in the near future optimize this function in GAPHYOR .

KEYWORDS

Scientific databases, productivity ratio, data integrity, efficiency ratio, production cost, bibliographic references, homonyms, truncation

REFERENCES

[1] Delcroix J.L., Manuel interne GAPHYOR, Orsay (1994)
[2] Delcroix J.L., Gaphyor Rept n°XX, Orsay (1994)
[3] Delcroix J.L., Humbert D., Leprince C., GAPHYOR HANDBOOK, Orsay (1993)
[4] Delcroix J.L. et al, GAPHYOR UPDATE vol 21, Orsay (1995)

PART TWO

RIGID AND FLEXIBLE SHAPES

REPRESENTATION AND PROCESSING

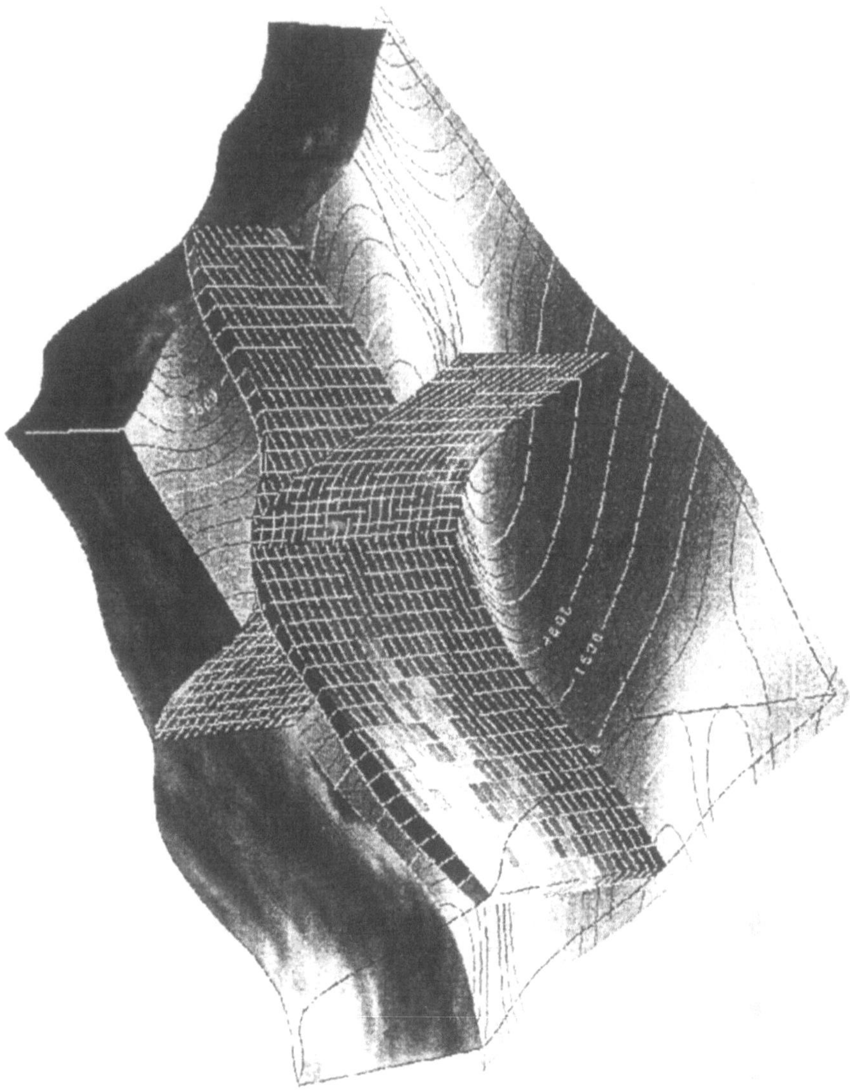

Representing the physical properties of natural 3D objects is a complex task. It involves specific computer structures and objects. A curvilinear stratigraphic grid is used to model the petrophysical parameters of this sedimentary oil bearing formation. The cells are more and less regular according to the thickness variations of the layers. Properties are estimated using a conditional gaussian simulation. These techniques are extensively used for reservoir modeling in the oil industry (Document GOCAD - after Shtuka A. et al.)

Chapter 4

SHAPE AND SURFACE CAO PROCESSING

SHAPE - DATA PROCESSING IN THE NATURAL SCIENCES AND TECHNOLOGY

Paul G. MEZEY

Molecular Modeling Group, Mathematical Chemistry Research Unit, Dept. Chemistry and Dept. Mathematics, University of Saskatchewan, 110 Science Place, Saskatoon, Canada, S7N 5C9.

ABSTRACT

The shapes of objects have major importance in science and technology. The analysis and comparison of objects, the detection and evaluation of their similarities and differences, understanding their function and behavior, testing theoretical models, and the design of new objects in engineering, all depend on shape. The analysis, quantification, numerical representation, organization, the storage, and retrieval of shape information of shape databases have become important components of the effort to manage the accumulated knowledge of mankind. In this study the general aspects are discussed, with special emphasis on molecular shapes, relevant to chemistry, biochemistry, pharmacology, medicine, and the related industries.

RESUME

La forme qu'ont les objets est extrèmement importante, tant en science qu'en technologie. De la forme dépendent notre analyse et notre comparaison des objets, la découverte et l'évaluation de leurs similarités et différences, notre compréhension de leur fonction et de leur comportement, le contrôle de nos modèles théoriques, et -en génie- le développement de nouveaux produits. Dans cette étude nous discuterons des aspects généraux de la question, tenant compte surtout des formes moléculaires, qui ont des conséquences en chimie, biochimie, pharmacologie, médecine et les industries qui en découlent.

1 INTRODUCTION

Shape is one of the most fundamental concepts in the natural sciences and technology. It is often tempting to approach a scientific or technological shape problem the way we evaluate the shape of objects around us, primarily by visual inspection. Yet, in our everyday life the shape concept is somewhat subjective and elusive: two observers are likely to judge the shapes of two objects differently; to each observer a different aspect of form may appear important. This leads to some ambiguities and to a lack of reproducibility of shape evaluations. In our technological age it is particularly important to base shape information on firm data and to use precise methods for their evaluation and interpretation. Consequently, the generation, processing, and evaluation of shape-data by reliable computer methods have acquired exceptional importance during the past decade.

The primary difficulty is the wealth of data required for conventional shape characterization of all but the simplest objects. However, it is possible to reduce the amount of data and to extract the essential shape information by systematic methods. These methods rely on a reformulation of some of the key aspects of visual evaluation in a precise and reproducible manner, leading to non-visual shape-data processing computer programs. In this contribution some of the modern methods of computer-based shape analysis, shape-data generation and shape-data processing are reviewed.

The computational utility of these methods are demonstrated by several examples.

2 A DIFFICULT SHAPE PROBLEM : SHAPES OF MOLECULES REPRESENTED AS DYNAMIC, QUANTUM MECHANICAL OBJECTS

Among all the difficult shape problems, the shape characterization of molecules is particularly hard. Molecules are quantum mechanical objects, subject to the Heisenberg uncertainty relation; molecules never "rest", motion and dynamic features are inseparable components in any model designed for a faithful characterization of their shapes. Molecules do not have a finite body defined by precise geometrical conditions; molecules have no finite boundary surface that encloses their entire electron density.

The cloud-like, fuzzy electronic density of a molecule does not behave like a macroscopic body; there is no finite distance beyond which the electronic density of the molecule is precisely equal to zero, hence no precise distance can be given where the molecule "ends". At best, the peripheral regions of a molecule can be described by a continuous, 3D electronic charge density function that approaches zero value at large distances from the molecular center of mass.

In an approximate sense, however, the concepts of a formal molecular body and molecular surface are very useful within both classical and approximate quantum chemical models of molecules.

The boundary surfaces of a sequence of formal molecular bodies at various electron density thresholds can be defined, and one may regard some of these surfaces as enclosing the "essential part" of the molecule. These simple, intuitive concepts of molecular body and molecular surface are often used for the modeling of molecular size and shape properties, and for the description of conformational changes and chemical reactions. Molecules can be regarded as three-dimensional objects which occupy some space. Some of the possible choices for shape representations, and the related topological shape analysis techniques are described in ref. [1]. If a series of electron density thresholds are chosen, then these techniques can be formulated in terms of contour surfaces of electronic charge densities, called *molecular isodensity contours* (MIDCO's). Among the alternative choices *molecular electrostatic potential contours,* (MEPCO's), *nuclear potential contours* (NUPCO's), the contours of individual molecular orbitals, for example frontier orbitals, simple fused sphere molecular *Van der Waals surfaces* (VDWS'), solvent accessible surfaces, and various other surfaces are prominent.

For example, a formal molecular body B(a) can be represented by the subset of the three-dimensional space enclosed by a MIDCO G(a) of a density threshold a.

Similarities in the reactivity of different molecules are often associated with local similarities in their MIDCO, NUPCO and MEPCO surfaces. This indicates the special importance of electronic and electrostatic interactions during the initial stages of chemical reactions. The shapes of these surfaces often suggest mechanistic explanations of chemical and biochemical process, for example, of how a given ligand interacts with a receptor site of an enzyme.

The study of the three-dimensional shapes of MEPCO's, NUPCO's, and MEPCO's is an important task in rational drug design.

Visual shape analysis methods and shape comparisons are often subjective and seldom reproducible. Nonvisual, algorithmic similarity analysis based on automated shape and similarity assessment by a computer is a more reliable approach. Such algorithmic shape and similarity analysis techniques, the determination, evaluation, and comparison of shapes are important in many fields of natural sciences and technology. The topological methods of shape analysis simplify the molecular problem by focusing on the essential features and by describing shape in terms of topological invariants, leading to a shape code.

Compared to the topological methods, the simple geometrical models of molecules are restricted: the quantum mechanical uncertainty, the conformational flexibility and other dynamic properties of molecules are not described properly.

Within the topological model, one considers a continuum of a family of molecular surfaces collectively representing the shape of the molecule. Within this family, topological equivalence classes are defined, and these classes are characterized by algebraic topological tools. For example, the homology and homotopy groups are efficient tools for the shape analysis of families of surfaces. These groups are suitable to extract the dominant, common features from an entire family of MIDCO, NUPCO, or MEPCO surfaces [1].There are two distinct stages of the general topological approach to molecular shape analysis.

The primary objects are the individual molecular surfaces and the molecular bodies enclosed by them (geometrical objects). Using their common topological properties, equivalence classes are generated, and the secondary objects are these classes, each represented by a topological object. The common topological properties are defined, for example, by intervals of possible curvature values or a range of electron density contour values.

The basis of this two-step process is the principle of *Geometrical Classification and Topological Characterization* [1] that, in turn, is based on the related principle of *Geometrical Similarity as Topological Equivalence,* referred to as the GSTE principle [1]. In all these approaches, geometrical conditions are used to define ranges of geometrical objects (for example, families of points along a MIDCO where the contour surface satisfies some local curvature condition): a *geometrical classification* of the surface into domains.

The next step is a *topological characterization* of the interrelations among these domains. The topological characterization is *stable* with respect to most small geometrical variations, and topology is used to extract the essential shape information that is modified only by greater geometry variations.

Various geometrical or physical conditions can be used. These conditions, denoted by μ, define various domains on a contour surface G(a), for example, the curvature based shape domains D_μ, where $\mu=0$, 1, and 2 correspond to locally concave, saddle type, and convex domains, respectively, relative to some reference curvature b. These domains subdivide the contour surface G(a). By eliminating from G(a) all domains of some specified properties μ, a new, topologically more interesting object, a truncated contour surface $G(a,\mu)$ is obtained. This surface $G(a,\mu)$ "inherits" the essential shape information of the original surface G(a), and represents it in a topologically accessible manner. Except for the trivial truncation, the truncated surface $G(a,\mu)$ is no longer topologically equivalent to G(a). The topological invariants of the family of G(a,μ) surfaces define the *shape groups* of the contour surface G(a), with respect to the given family μ of shape domains. The shape groups are the *homology groups* of the truncated contour surfaces G(a,μ). The

Shape Group Method (SGM) [1] is a topological technique for a concise representation of the interrelations among the shape domains of molecules. The shape group method combines some of the advantages of geometry and topology, providing a symmetry-independent characterization of molecular shape.

The above implementation of the shape group method relies on a classification of surface domains based on local curvature properties. Curvature can be defined as the second derivative of a function describing the surface. The curvature-based methods of shape analysis require that the surfaces are twice differentiable. Some simple models of molecular surfaces do not fulfill this criterion; for example, fused sphere Van der Waals surfaces do not possess first derivatives at the seams of interpenetrating spheres, and dot representations of solvent accessible surfaces are not even continuous.

Various methods of shape analysis are applicable for nondifferentiable and noncontinuous surfaces [1]. One such technique is the *the method of* T-*hulls* [1], based on a generalization of the concept of *convex hull.* The *convex hull* of a set A is defined as the smallest convex set that contains A. An alternative, equivalent definition of convex hull is the basis of an interesting generalization. We shall use the following terminology: a half space is one side of the three-dimensional space divided by a plane. In order to re-define the convex hull of A, take all possible half spaces that contain the set A, and define the convex hull of A as the intersection of all these half spaces. A generalization of this approach is obtained if one replaces the half spaces by some other object T.

For example, consider a three-dimensional body T. The T-hull of the point set A is defined as the intersection of all rotated and translated versions of object T which contain A.

The T-hull method is applicable for shape comparisons of molecules, when a common reference shape is specified. Take this reference shape as that of the body T. The shapes of two molecules, T and A, can also be compared directly. The T-hull of the molecular body A provides a direct shape comparison [1]. A small deviation between the volumes of A and the T-hull of A indicates a high level of compatibility between T and A.

The *resolution based similarity measures* (RBSM) [1] form another important class of methods used in molecular shape-similarity analysis. The basis of these methods can be illustrated by a simple example. Assume that three objects, A, B, and C, appear indistinguishable at a great distance to an observer. As the observer approaches the objects, at some distance one of the objects, for example, object C, becomes distinguishable from objects A and B, but A may still appear indistinguishable from B. At a close distance, the observer can distinguish A from B. One may conclude that A is more similar to B, than C is to either A or B. Closer approach can be replaced by higher resolution. Using varying levels of resolution, we conclude that A and B are more similar to each other than A is to C or B is to C, if the dissimilarity of C from the other two objects is already evident at a medium resolution, whereas it takes a much higher resolution to distinguish A from B. If the shapes of two objects are identical, then they are indistinguishable even at infinite resolution.

One can define a *similarity measure* based on the level of resolution required to distinguish objects. This shape-similarity measure is called a *Resolution Based Similarity Measure* (RBSM) [1].

Resolution and distinguishability can also be represented within a general topological framework. If in a family of objects a common hierarchy of topologies is defined for each object, and if this hierarchy is ordered by the finer-cruder relations of the topologies (see, e.g. ref. [1]), then considering finer topologies is analogous to considering higher levels of resolution. This leads to Topological *Resolution Based Similarity Measures* (TRBSM's).

Molecular shapes cannot be described realistically without taking into account the *dynamic aspects* of their motion, their quantized vibration, the large amplitude displacements of various parts of the molecules relative to one another, and their conformational changes[1].

Several rules have been described, interrelating the mechanism and outcome of conformational changes, electronic excitations, and chemical reactions, placed within the common framework of the energy dependence of the accessible shapes and the accessible symmetries of various molecules [1-3].

For small molecules the standard, *ab initio* quantum chemical computations provide reasonably accurate representations of molecular shapes in terms of approximate electron densities. However, electron density computations of similar accuracy for large molecules were realized only recently. A simple principle, the additive fuzzy density fragment principle has been shown to be exact for molecules [4]. Based on this principle, the Mulliken-Mezey fragmentation scheme, and a set of topological rules, a new computational method, the Molecular Electron Density Lego Assembler (MEDLA) technique was proposed[4]. This technique is suitable for the construction of *ab initio* quality, yet inexpensive electron densities of large molecules [4].

The MEDLA method constructs approximate electron densities by utilizing an interpenetration pattern of pre-calculated, *ab initio* electron densities of fuzzy molecular fragments. These fragments are obtained from direct, high quality *ab initio* calculations on small molecules and stored in a *fragment density data bank.* These density fragments are realistic, fuzzy electron distributions; they have no boundaries.

The reconstructed MEDLA electron density is exact (within the given *ab initio* framework) if the fragments originate from the given molecule. Non-exact, but reasonably accurate, *ab initio* quality densities can be obtained for other molecules, including large biomolecules, built from such fragments.

The computation time needed for the calculation of MEDLA densities grows (essentially) linearly with molecular size. A modest data bank of less than two hundred fragment types is sufficient to build electron densities for many biomolecules. The fragment densities can be retrieved from the data bank and used by the computer Program MEDLA 93 [5], to "build" electronic densities for various molecules, such as the first high-resolution electron density images of bovine insulin[7], shown in Figure 1.

This method allows one to extend detailed shape analysis methods and density domain rules [6] for large molecules.

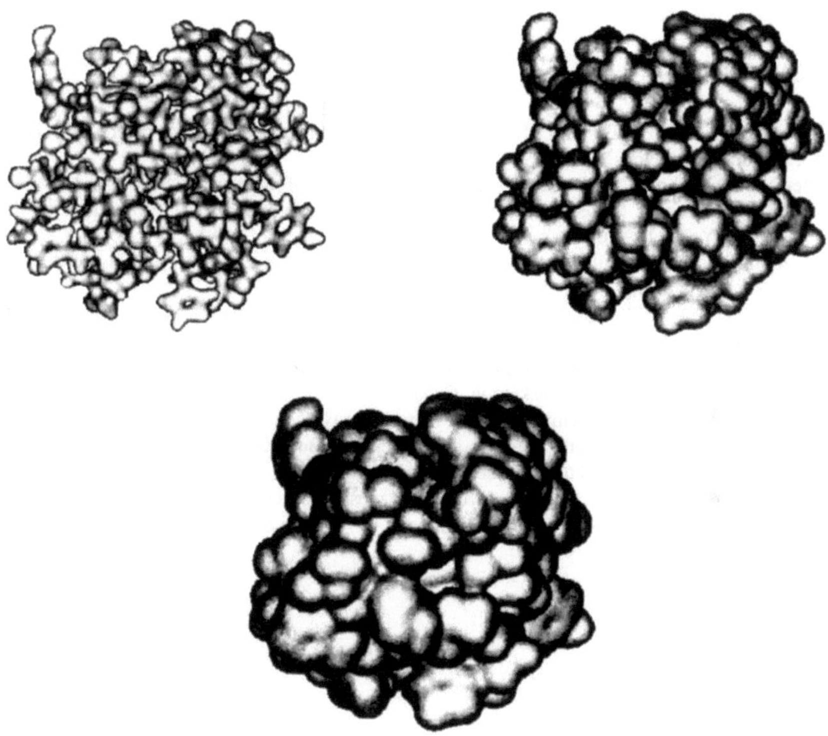

Figure 1: The first high-resolution electron density images of bovine insuline, calculated at density thresholds of 0.1, 0.01, 0.001 atomic units, using the MEDLA technique [7]

CONCLUSION

Some of the topological techniques of shape data generation and processing have been reviewed, with special emphasis on molecular shapes.

ACKNOWLEDGMENT

The original development of topological shape analysis techniques and the MEDLA method was supported by both strategic and operating research grants from the Natural Sciences and Engineering Research Council of Canada.

KEYWORDS

Shape analysis, shape databases, object design

REFERENCES

[1] Mezey P.G., SHAPE IN CHEMISTRY : An Introduction to Molecular Shape and Topology, VCH Publishers, New York (1993).

[2] Mezey P.G., Three-Dimensional Topological Aspects of Molecular Similarity, in Concepts and Applications of Molecular Similarity, pp 321-368, Johnson M.A., Maggiora G.M. (eds) Wiley, NewYork (1990).

[3] Mezey P.G., Shape-Similarity Measures for Molecular Bodies : A 3D Topological Approach to QShAR., J. Chem. Inf. Comp. Sci., 32, 650-656 (1992).

[4] Walker P.D., Mezey P.G., Molecular Electron Density Lego Approach to Molecule Building, J. Amer. Chem. Soc., 115, 12423-12430 (1993).

[5] Walker, P.D., Mezey P.G., Program MEDLA 93, Mathematical Chemistry Research Unit, University of Saskatchewan, Saskatoon, Canada.

[6] Mezey P.G., Quantum Chemical Shape: New Density Domain Relations for the Topology of Molecular Bodies, Functional Groups, and Chemical Bonding, Canad. J. Chem., 72, 928-935 (1994). (Volume dedicated to Professor Polanyi).

[7] Walker P.D., Mezey P.G., Realistic, Detailed Images of Proteins and Tertiary Structure Elements: Ab Initio Quality Electron Density Calculations for Bovine Insulin, Canad. J. Chem., 72, 2531-2536 (1994)

3D Modeling Of Complex Natural Objects

Jean-Jacques ROYER, Benoit GERARD, Christian Le CARLIER de VESLUD and Arben SHTUKA

C.R.P.G., B.P. 20, 54501 Vandoeuvre-lès-Nancy Cedex, France

Abstract

The aim of 3D modeling is to define the geometry and properties (such as permeability, porosity,...) of natural complex structures which are investigated from irregular sampling (boreholes, sections, points). Once a geometrical model is built, the evolution of various parameters, such as coupled heat and mass transfers, can be studied using numerical modeling. The geometry and the physical parameters of the objects are modeled using the GOCAD software developed at the Ecole de Géologie de Nancy. Several studies are presented to illustrate this approach, including the 3D modeling of an underground chromite mine, the thermal field and the fluid flow patterns around a fractured geothermal site at Soultz in the Rhine Graben, and the 3D seismic data collected during an injection experiment.

Resume

Le but de la modélisation 3D est de définir la géométrie et les propriétés des objets naturels complexes à partir d'informations partielles, irrégulièrement réparties dans l'espace (sondages, sections, échantillons,...). Un modèle géométrique est construit, les différentes propriétés sont calculées par modélisation numérique (équations couplées chaleur-masse, écoulements fluides,...). La géométrie et les paramètres physiques sont modélisés à l'aide du logiciel GOCAD, développé à l' Ecole de Géologie de Nancy. Plusieurs études de cas illustrent cette méthodologie : la modélisation 3D d'une mine souterraine de chromite, le champ de températures et les écoulements fluides autour du site géothermique de Soultz (Graben du Rhin), ainsi que la sismicité observée en 3D dans ce site lors d'un test d'injection.

1 Introduction

3D modeling of complex geometrical objects is of primary importance for forecasting the properties of objects from a reduced number of parameters and observations. Methods for 3D modeling can be separated into two categories: namely, *CAD methods* and *natural methods*. In the CAD approach, the topology of the object is perfectly controlled at each step of the conception and of the elaboration.

This approach mainly concerns manufactured or industrial objects which might be of complex shape. Most of the time, triangular, tetrahedral or quadrilateral elementary elements are used to fit the surface or the volume of the final object using spline or Bezier interpolation methods. Several industrial CAD systems based on such concepts are widely used in the industry; among the most popular are AutoCAD and STRIM 100. It is difficult to apply this approach to natural objects because the topology of the underlying real object is not known, furthermore, natural objects are usually investigated from irregular sampling (boreholes, sections, points). This uncertainty as to real shape implies that the classical CAD approach might be irrelevant, as shown by several pioneer works [9].

In fields such as geology or environmental sciences, one must frequently decipher the underlying structure from indirect observation or from scarce, irregular, or partial data points. The situation can be further complicated by the presence of discontinuities such as faults, fractures or folded layers. This is why the natural object approach has been developed specifically to handle natural objects.

In this paper the geometry and the physical parameters of objects are modeled using the GOCAD software. Once a geometrical model is built, the evolution of various parameters can be studied using a numerical interpolation method or by resolving the equations governing the underlying physical processes. This approach is illustrated by several examples.

2 Presentation of the GOCAD Software

The GOCAD software, developed at the Geological School of Nancy, is a powerful tool designed for interactive 3D modeling of surfaces and volumes, including geophysical applications such as 3D ray tracing, 3D tomography and migration and reservoir simulation of complex geological structures (reverse faults, salt domes,...). It comprises a Geometrical Data Base specially designed to efficiently carry out operations such as

- intersection or union of two or more surfaces,
- intersection of a surface by lines (ray tracing), surfaces or volumes,
- storage and interpolation of physical properties attached to a layer (velocity, density, porosity, permeability,..) or to a volume.

The visualization of the information contained in the data base is done interactively using the X-Window/Motif interface under the Unix operating system. The code, written is C and C++ in an object oriented style, includes a device-independent 3D graphic library allowing it to run currently on Convex, Digital, HP, IBM, Silicon Graphics and SUN computers. (Additional information about this system can be obtained from the GOCAD project managed by the Association Scientifique pour la

Géologie et ses Applications, Pr. J.L. Mallet, Chairman of the Computer Sciences Department of the Nancy School of Geology, Nancy, France). In this approach, surfaces are modeled as two dimensional graphs whose node locations are determined from different types of heterogeneous data using the Discrete Smooth Interpolation DSI method [9]. The same DSI method can also be used to interpolate any value within the domain (properties or parameters).

3 MODELING OBJECTS IN 3D

Modeling the geometry and the physical parameters of an object using the GOCAD software usually involves four different steps (see Figure 1) :

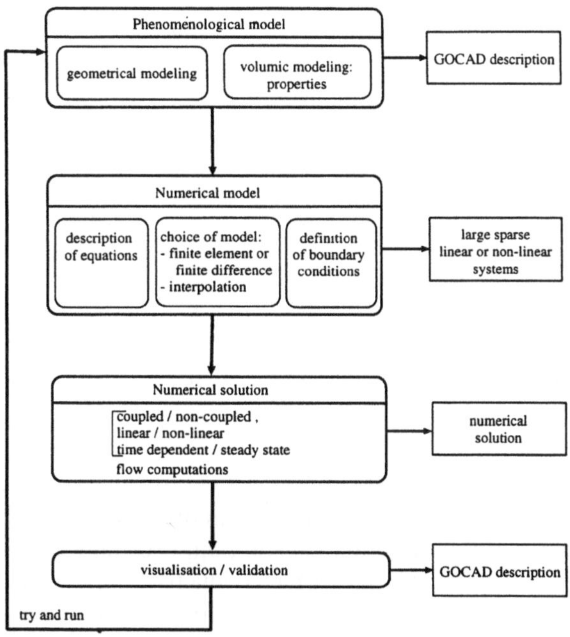

Figure 1. Schematic representation of modeling process.

1) *Building the geometrical model* : First, the geometry is modeled using surfaces defining the limits of the studied object (layers, volume,...), the limits of the domains (boundaries), and the discontinuities (fractures, faults). Then, each volume (volume modeling) is gridded into a set of 3D elementary elements on which the properties of different domains are defined. These steps involve the interpolation of surfaces and properties of the objects (permeability, Young modulus, thermal conductivity,..) into the space from a given set of "control points" defined at node locations at which several parameters are observed. This information can be "hard"

(quantitative), but it can also be uncertain or "soft" (qualitative). The interpolation of properties of the objects are made into the 3D space using the DSI method.

2) *Modeling the equations* : In the second step, the set of partial differential equations involving unknowns (such as temperature field or filtration velocity) and properties (such as Young modulus or tensor of permeability) governing the phenomena is defined. For example, in the case of fluid flow study in heterogeneous and anisotropic media, four types of equation have been investigated [12]. They refer to coupled, linear or non linear differential equations, among them the heat equation, fluid flow in porous medium, mass balance and behavior laws of rock properties, defined for the different materials as follows :

- *Fourier law* :

$$-\nabla . ([\Lambda]\nabla T) + \rho C \frac{\partial T}{\partial t} + \rho_{fluid} C_{fluid} \mathbf{v} . \nabla T = A \quad (1)$$

- *Darcy law in porous medium* :

$$\eta \mathbf{v} + [\mathbf{K}] (\nabla P + \rho_{fluid} g \mathbf{e}_z) = \mathbf{0} \quad (2)$$

- *Mass balance* :

$$\nabla . (\rho_{fluid} \mathbf{v}) = 0 \quad (3)$$

- *Behavior law for properties* :

$$\lambda(T) = \frac{\lambda_{ref}}{1 + \alpha(T - T_{ref})} \qquad \rho_{fluid}(T) = \rho_{ref} (1 - \beta_{th} (T - T_{ref})) \quad (4)$$

where:

A = heat production, e_Z = unit vector for the coordinate system along the vertical; g = acceleration of gravity; [K] = tensor of permeability; P = pressure; t = time; T = temperature; v = Darcy filtration velocity; b_{th} = volume expansion coefficient of the saturating fluid; [L] = tensor of thermal conductivity; h = dynamic viscosity of the saturating fluid; rC = specific heat content (rock + fluid) at constant pressure; r_{fluid} = mass density of the fluid.

3) *Numerical resolution and handling* : The above partial differential equations are discretized using a finite difference or finite element mesh in order to transform them into a set of coupled non-linear equations, but they are linearized during the solving. Boundary conditions are then introduced, making the system non-singular. The resulting linearized system leads to the handling of large and sparse matrices. This is one of the key points of the method because 3D modeling involves a great number of unknowns and subsequently, the corresponding linear systems are huge and sparse. It is thus necessary to store only the relevant part of information using a special sparse data structure using pointers. Another key point is the

resolution of the linear systems. It is necessary to solve large systems (125 000 x 125 000 is common) as fast as possible, but also to ensure numerical robustness of the solution to avoid oscillations or non-convergence. The following methods have been implemented and tested: (i) Gauss-Seidel: a top-hits classical robust method, but slow to converge [11]; (ii) conjugate gradients: a more efficient method, but not convergent for all kinds of matrices [11]; (iii) full multigrid techniques: the fastest and to-date most promising method, but a difficult one which requires delicacy in implementation [5]. Because of the huge computer resources necessary for solving such a problem, this step is usually done in batch mode.

4) *Visualization and validation of the model* : The visualization is done interactively using the GOCAD software.

4 ASSOCIATED DATA STRUCTURES

Specific data structures have been added to the current GOCAD ones in order to manage the information necessary in the solving procedure. They are briefly described in Figures 2 and 3.

We introduce some definitions (see Figure 2) :

- *support* : a part of 3D space.
- *element* : the basic unit of the model. An element is a set of points associated with a simple geometry. It must belong to only one domain.
- *domain* : a set of elements which share the same behavior.
- *boundary condition* : condition imposed at the boundary of the support for describing the exchanges with the remainder of 3D space.

These definitions are used by the following structures to handle the solving process.

- The *problem* structure contains the geometry of the studied area, the definition of domains and the property of each domain. The equation and the variable lists are defined to handle several coupled equations describing one unknown per equation.
- The eq*uation* structure stores both numerical information (type, boundary conditions, linear system) and geometrical information. It is linked to a set of pointers to macro functions whose purpose is to simplify the manipulation of properties, equations and unknowns. This programming technique supports in a simple way all kinds of different situations such as parameters depending on time, space or upon each other (for instance temperature, pressure,..). A general flow chart shown in Figure 2 summarizes the links between the structures defined above.

- The *boundary condition* structure contains the type of the boundary conditions such as imposed potential (Dirichlet) or imposed flow rate (Neumann). It refers to a surface to which the boundary condition is applied.
- The *property* structure contains the nature of the property which is defined for each element . It can be :
- interpolated from regular or irregular sample points at control nodes
- simulated at each node using geostatistical methods in order to take account of the spatial variability of parameters.
- The *behavior* refers to the evolution of properties through time or space. It can be constant or variable depending on the state variables such as temperature, pressure or straightness.

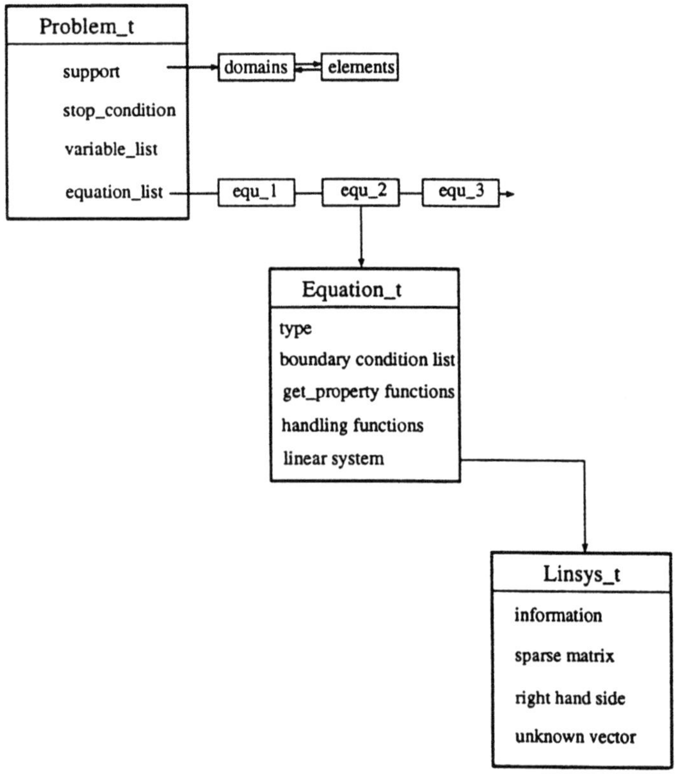

Figure 2. Basic elements used in 3D modeling.

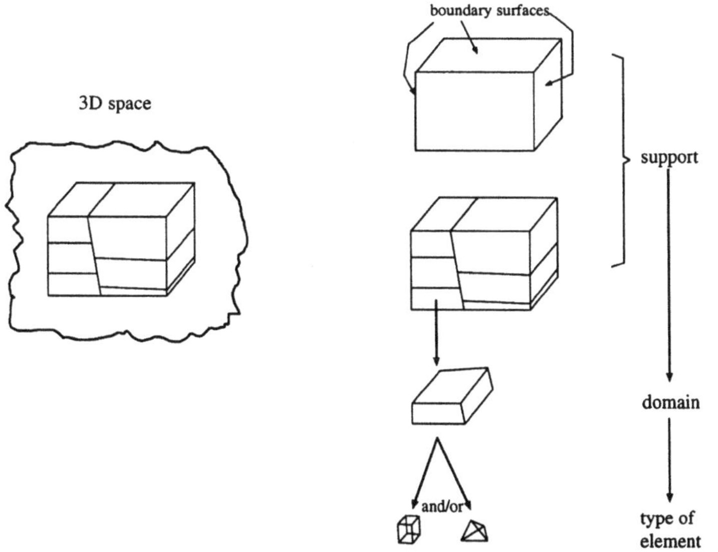

Figure 3. Schematic representation of some related structures.

5 EXAMPLES

In order to illustrate the different steps required in the modeling process, four examples have been selected :

1) *Reconstruction from borehole and gallery sampling of the shape in 3D of a complex faulted stratiform chromite deposit.* This deposit is located in a ophiolitic belt. Two type of chromite deposits are defined, depending on their chemical composition: (i) subcordant Cr deposits, high in grades and situated at deep stratiform level; (ii) discordant Cr deposits situated nearer to the surface with Al and Ti. They are both situated in fractured tectonized zones leading to complex deformed and fractured ore deposits. The mineralization is thought to be of magmatic and/or post-magmatic hydrothermal origin involving interaction between the magma and the magmatic fluids. The deposit studied is exploited on 9 levels from which 2500 samples have been recorded from the different exploitation boreholes. Because of the complexity of the regional tectonics, it is difficult to represent the geometry of the mineralization in 3D space. The original data obtained from borehole survey is the location and depth of the top and bottom of the mineralized layers. In addition, the observed surface or underground fractures are added to the geometrical model as surface discontinuities [15]. The final geometrical model takes account of the different tectonic phases which have restructured the deposit, such as faults, discontinuity, folding (Figure 4).

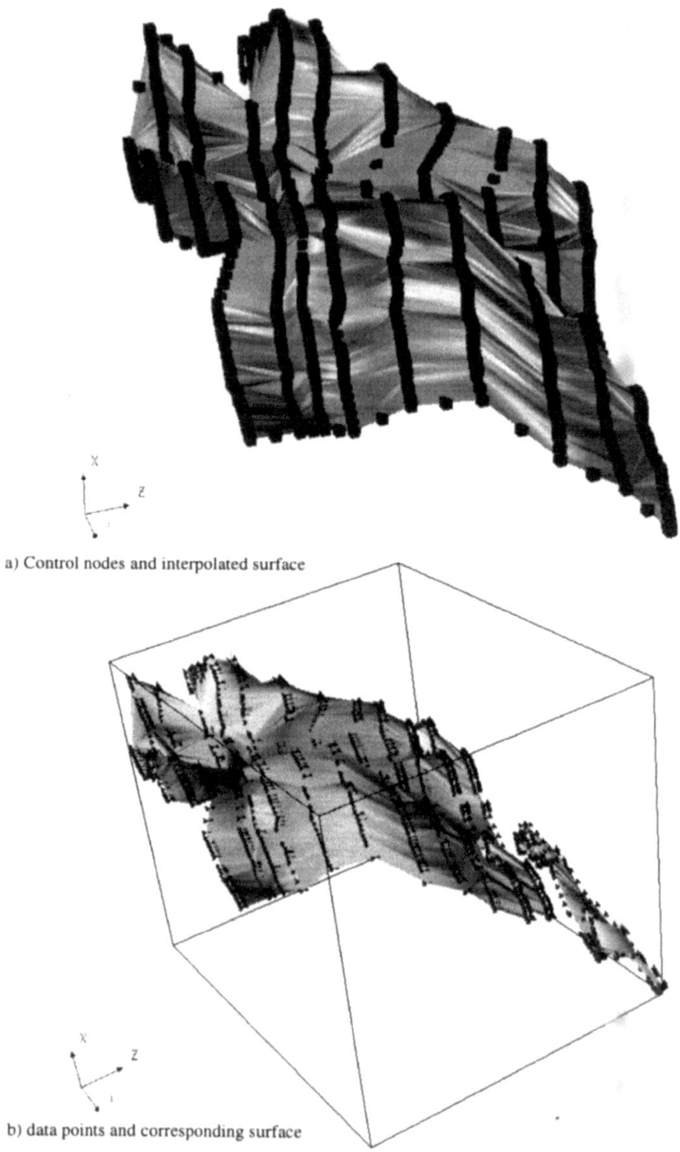

Figure 4. The chromite mine case study: 3D view of the top of the mineralized layer. a) Representation of the control nodes (nodes where surface is known) and the interpolated surface, numbering 4411 nodes and 7943 triangles. b) A more complete 3D view of data points and the corresponding surface.

2) *3D geothermal modeling of a geological sedimentary formation at Soultz in the Rhine Graben.* Located at the western limit of the upper Rhine Graben, Soultz-s-Forêts is a pilot zone selected by the European Hot Dry Rock (HDR) project whose purpose is to exploit geothermal low enthalpy energy [1] [7] [14]. It is of primary importance to evaluate the potentiality and the present thermal state of the region before exploitation. For this reason a research project has been launched on the modeling of heat, fluid and mass transfer in the Soultz region. As a preliminary step, the geometrical model of the underlying sedimentary layers and of the granitic bedrock including the major faults has been made by the BRGM, using vertical seismic cross-sections and the oil wells available on the site (more than 45). During the modeling, the seismic data was reinterpreted in order to provide a coherent 3D set of faults. This model is then used to assign the physical properties (thermal conductivity, permeability, porosity,...) to each facies rock (domains) [6] [13]. The faults are considered as highly anisotropic porous media. Once the physical and geometrical model has been made, the coupled transfer equations (heat, Darcy and mass balance) can be solved, assuming different boundary conditions [3] [2]. The results show clearly the presence of convective cells under the Soultz site causing hydrothermal circulations and thermal anomalies (Figure 5).

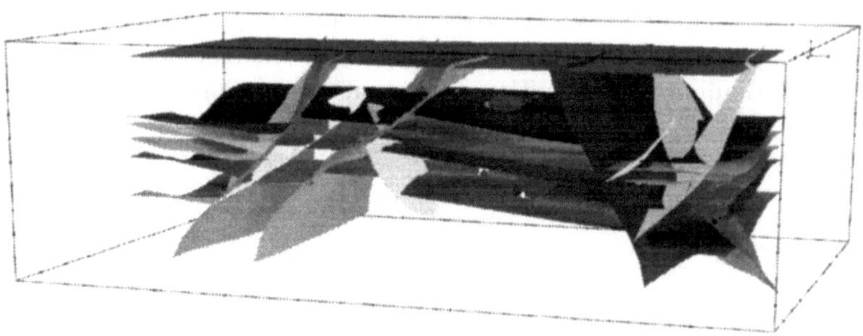

a) 3D model of the Soultz region

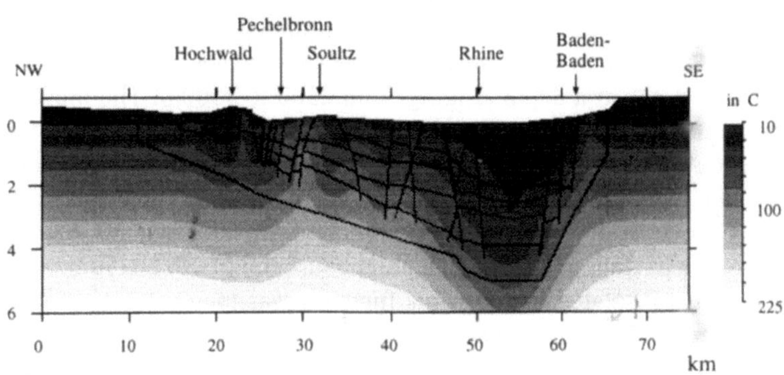

b) Calculated temperature field along a vertical cross section

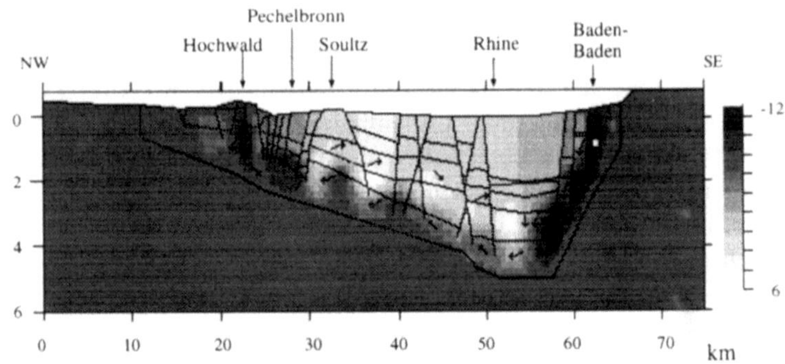

c) Calculated stream function along a vertical cross section.
Arrows show the path of the fluid.

Figure 5. Modeling temperature and fluid transfers at Soultz (the Rhine Graben). a) 3D representation of the sedimentary layers and faults of the geothermal site. This geometrical model is used to obtain the geology along any cross-section. b) Temperature field along a NW-SE cross section. c) Stream function along the same cross-section. The arrows show the path of the fluid. The temperature field is strongly influenced by fluid circulations. The circulations can explain geothermal anomalies discovered at Soultz.

3) *Representation in 3D space of the seismic cloud during an injection test in a fractured granitic bedrock.* The seismic data were recorded by the Camborn School of Mines during an injection test performed in the GPK1 well situated at Soultz. The goal of the study was to characterize the density and the orientation of the faults of the granitic bedrock at a depth ranging from 2500 to 3500m. The method records the coordinates and the time of each crack observed in the medium during fluid injection. The 3D representation of such data is not easy because they are numerous (more than 20 000) and irregularly spread in space. In order to obtain the envelope of the fracturation zone, a density function was computed at each 3D grid sample point using a moving average method and a neighbourhood of 50x50x50m. A threshold was used to eliminate isolated events. This function was then interpolated by DSI onto a regular 3D grid and finally visualized using the GOCAD software. The results can be represented by isodensity surfaces, depicting the distribution of crack events in 3D space, or as a function of time, depicting the spreading of the cracks (Figure 6).

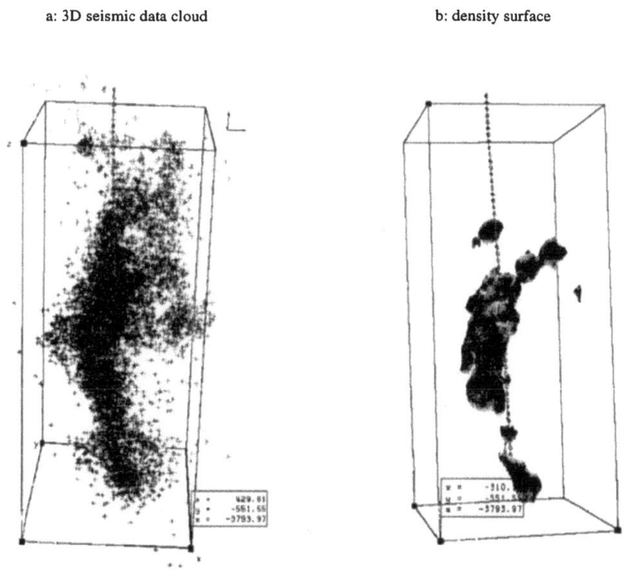

Figure 6. Representation of a seismic cloud during an injection test in a fractured granitic bedrock. a) The seismic data, numerous (more than 20 000) and irregularly spread. b) Density function isovalue, after interpolation on a regular grid. Preferential orientations of the cloud appear more clearly.

4) *Modeling of a buried natural fission reactor at Oklo, Gabon.* The Francevillian uranium deposit at Oklo is unique because it hosts the only natural fission reactor known to have occurred [10]. The fission reactions occurred two billion years ago. The reaction zones in the Oklo area are similar to a burned-up nuclear reactor and represent the only natural example of fission products stored in a geological environment for a geological period of time. Understanding the impact of the buried reactor involves understanding the thermal effect of high heat production zones (from 10 to 150 Wm^{-3}) on surrounding host rocks, and consequently the influence of this heat on the fluid movements and migration of matter. To do so, a geometrical model was first investigated from mine plans, drilling data and indirect geological considerations. Then, the physical properties of the different geological formations were assigned according to temperature dependence, leading to a non-linear system for the coupled transfer equations. The results shown in Figure 7, show clearly that the reactor behaved like a powerful local perturbation of the thermal regime. During the fission reaction, the temperature at the center of the reactor increased by 50 to 150 °C above the regional one, depending on the assumed permeability and heat production, while the thermal perturbation around the center of the reactor extended to less than 50m [8] [4]. The filtration fluid velocities were low

(about 10^{-8} to 10^{-10} m.s^{-1}) compatible with a conductive thermal regime. In this case, the functioning of the reactor did not modify significantly the mass balance budget and the circulation of the water through the sedimentary layers at a regional scale.

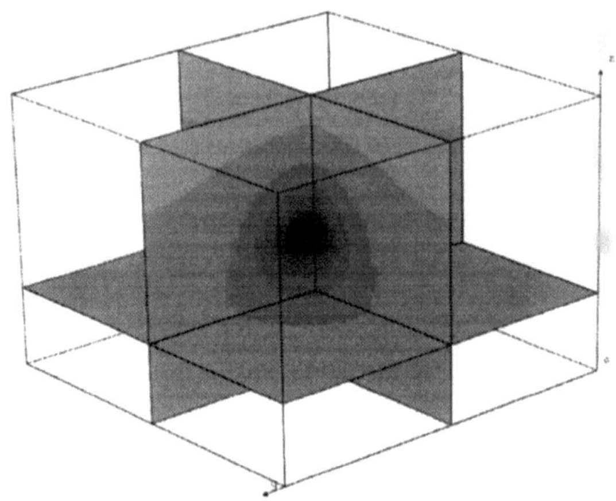

a) 3D Temperature field around the reactor, assuming a heat production of 100 W/m³ (conductive case)

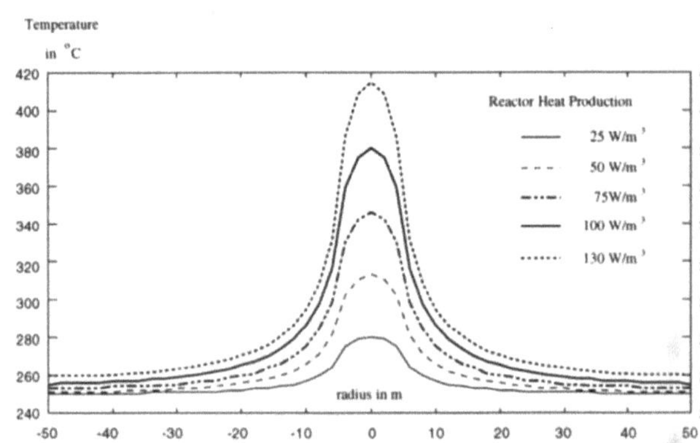

b) Heat Production Influence: temperature vs distance from the reactor

Figure 7. Thermal study of a natural fission reactor in Oklo (Gabon). The purpose is to know the influence of a buried reactor on surrounding rocks. a) 3D representation of the temperature field, assuming a heat production of 100 Wm^{-3}. The reactor is a thin parallelepiped situated at the center of the cube.b) Temperature profiles along a radius for various heat productions. The thermal influence of the reactor extends less than 50m, whilst the temperature at the center of the reactor increases by 50 to 150 °C above the regional one.

CONCLUSION : FURTHER DEVELOPMENTS

Modeling complex natural objects appears clearly as a important challenge. Significant results have already been obtained [9], nevertheless, some problems remain difficult to solve, concerning the data (managing huge amounts of data, of different kinds, with different scales); the theoretical and numerical processes (huge linear and non-linear systems, long computation times, equation instability, multiphasic flow). It should also be noted that natural objects study generally involves interdependent phenomena. For example, in oil engineering, the modeling of multiphasic flows coupled with heat transfers have to be used to determine position of source rocks, migration of oil and finally, entrapment in faulted zones. Similarly, modeling of coupled heat and mass transfer associated with dissolution/precipitation laws is needed to simulate ore deposit history. Thus, automatic handling for solving of complex coupled systems has to be implemented.

Several directions are under development: new methods of interpolation and of description of data taking into account the history and the logic of the studied bodies (for instance, tectonic history, compaction laws, deposit directions,...), implementation of efficient numerical methods (for instance, the full multigrid method for solving systems) and special formulation for multiphasic flows.

ACKNOWLEDGMENTS

This work was supported by the CNRS-PIRSEM Project. The data necessary for the representation of the geothermal Soultz site were provided by BRGM. The work on the Oklo reactor was supported by a CREGU-CEE project. The authors are indebted to SOCOMINE, especially to its director, A. Gérard for his general support, and to the CSM Associates Ltd, Rosemanowes Penryn University, for providing the 3D tomography data of the Soultz site.

KEYWORDS

3D modelling, natural objects, GOCAD, heat and mass transfers

REFERENCES

[1] Clauser C., Villinger H., Analysis of conductive and convective heat transfer in a sedimentary basin demonstration for the Rhine Graben, Geophys. J. Int., Vol 100, 393-414 (1990).

[2] Flores L., Transferts de chaleur et de masse en milieu sédimentaire et fracturé. Modélisation numérique de la convection naturelle autour du site de Soultz (Graben du Rhin) (1992).

[3] Flores L., Royer J.J., Convective heat transfer around the Soultz-s-Forêts geothermal site (Rhine Graben), Doc BRGM, Vol. 223, 155-170 (1993).

[4] Gerard B., Le Carlier de Veslud C., Royer J.J., Lavest P., Modélisation 3D Équations de transferts dans les corps sédimentaires, in 15ème Réunion des Sciences de la Terre, Nancy (1994).

[5] Hackbush W., Trottenberg U., Multigrid methods: Proceeding, Koln-Porz (1981).

[6] Jung R., Hydraulic fracturation and hydraulic testing in the granitic section of borehole GPK1 Soultz-sous-Forêts, Geotherm. Sci. & Tech., Vol 3, 149-198 (1991).

[7] Kappelmeyer O., Gerard A., Scholmer W., Ferrandes R., Rummel F., Benderitter Y., European HDR project at Soultz-sous-Forêts general presentation, Geotherm. Sci. & Tech., Vol. 3, 1-4 (1991).

[8] Le Carlier de Veslud C., Royer J.J., Savary V., Pagel M., Modélisation des transferts de chaleur et de masse lors des réactions naturelles de fission dans le gisement d'uranium d'Oklo (Gabon), in 15ème Reunion des Sciences de la Terre, Nancy (1994).

[9] Mallet J.L., Discrete Smooth Interpolation, ACM-TOG, Vol. 8, 121-144 (1989).

[10] Naudet R., Oklo: des réacteurs nucléaires fossiles, Etude physique, Paris (1981).

[11] Press W.H., Teukolsky S.A., Vetterling W.T., Flannery B.P., Numerical Recipes in C, 2nd ed, Cambridge (1992).

[12] Royer J.J., Flores L., Two-dimensional natural convection in an anisotropic and heterogeneous porous medium with internal heat generation, International journal of Heat and Mass transfer, Vol. 37, 1387-1398 (1994).

[13] Rummel F., Physical properties of the rock in the granitic section of borehole GPK1 Soultz-sous-Forêts, Geotherm. Sci. & Tech., Vol 3, 199-216 (1991).

[14] Schellschmidt R., Schultz R., Hydrogeothermic studies in the Hot Dry Rock project at Soultz-sous-Forêts, Geotherm. Sci. & Tech., Vol 3, 217-238 (1991).

[15] Shtuka A., Royer J.J., Geometric modelling and estimation of mineral deposits with complex shape, in Third CODATA conference on geomathematics and geostatistics, Enschede, Sciences de la Terre série informatique (1994).

QUANTITATIVE DESCRIPTION OF THE SHAPE OF INTERFACES IN MOLECULAR COMPLEXES

Robert BYWATER

Biostructure Department, Novo Nordisk A/S, DK-2880 BAGSVAERD, Denmark.

ABSTRACT

The Fourier shape descriptor method for describing complex shapes has become an established tool in molecular shape analysis [6 - 9]. An interesting further development of this technique is its use in describing and comparing matching surfaces between molecules in molecular complexes. In the case of biological systems the biological function is dependent on very intimate contact between the molecules that participate in the complex. Examples will be given of cases where small ligands are bound to protein binding sites, and of protein-protein complexes. Not only can the contact surfaces be accurately described, but it is possible to automatically dock one partner in the complex to the other.

RESUME

La méthode fondée sur les descripteurs Fourier de formes est devenue un outil reconnu de l'analyse des formes moléculaires [6 - 9]. Un développement intéressant ultérieur permet son utilisation pour la description et la comparaison des surfaces intervenant aux interfaces de molécules engagés dans des complexes. Pour les systèmes biologiques, la fonction biologique dépend du contact étroit et intime entre les molécules participant au complexe. Sont présentés des exemples où de petits groupes chimiques sont associés aux sites liants de protéines, et aussi à ceux de complexes protéine- protéine. Les surfaces de contact peuvent être décrites de façon précise ; de plus, on peut "insérer" de manière automatique un partenaire dans le complexe.

1 INTRODUCTION

Organic chemistry has concerned itself mainly with the structure and properties of covalently-linked molecular entities, and only relatively recently has there emerged an interest in the properties of molecular complexes, held together by weaker forces. In contrast to this, biochemistry and molecular biology are concerned not only with the covalently-linked molecules themselves but, to a very great extent, with the properties of molecular complexes. This is by no means a new development; the concept of a *receptor* which binds, noncovalently, a specific ligand is almost exactly a century old [1]. With the recent developments of computational chemistry and 3D imaging this receptor concept and its implications can be revisited more efficiently.

Molecular structures are conventionally represented as graphs in which each covalent bond is represented as a line joining two points corresponding to the positions of the atoms that are linked.

More information and realism, for example stereochemistry, can be incorporated into these models by drawing the bonds in a distinctive way as to whether they are in or out of the plane of the 2-dimensional drawing, or by drawing stereographic pictures.

Another conventional representation of molecular structure involves assigning a 'volume' to each atom by drawing a sphere centered at the nucleus. The radius of the sphere can be set at any desired value, depending on what feature of the molecule is being represented, but very commonly this will be the Van der Waals radius of the constituent atoms (standard values obtained from crystal structure databases). This representation of the molecule conveys qualitative information about molecular volume and space filling, but it is not useful for quantitative studies of, for example, the packing of molecules in molecular complexes.

Molecular surfaces

For displaying, describing and comparing molecular *surfaces,* a variety of techniques were developed earlier [2 -5]. More recently, the shape descriptor method has been shown [6 - 9] to work well as a means of quantifying shape, of measuring shape similarity/difference. One way of validating the method is to reconstitute the original molecule from its shape descriptors. The resulting structures have been shown [6] to be indistinguishable from the original. Shape descriptor data has furthermore been used as a variable, and found in many cases to be the dominant one [10], for use in quantitative structure/activity relationships where various structural parameters are invoked to account for a given type of activity, and their relative contributions as determined by statistical methods.

Biomolecules Interfaces

In this paper, the focus will be on describing the shape of the interfaces between biomolecules that are in intimate contact with one another. Before proceeding with this however, it is necessary to clarify exactly which 'surface' we intend to use as the starting point for these calculations. There is of course no hard-and-fast (non-fuzzy) definition of molecular surface, but there are a number of working definitions that can be employed, depending on what it is one wants to achieve. One example is the Van der Waals surface, referred to above. A more convenient version of this is the smoothed Van der Waals surface calculated using the **MS** program of Connolly [11]. It is this surface that has been used in the work reported here.

2 METHODS AND RESULTS

It has been shown [12] that it is just as easy to calculate the shape descriptors for a surface that is not entirely closed, such as the binding region on a macromolecule.

If each of the mutual binding regions on two molecules that are bound together are studied, they will both have the same, or very similar, shape descriptors, *provided that the original coordinates used to calculate the shape descriptors are those of the molecules in the docked position.* If the shape descriptors for these regions are calculated from the binding surfaces that are located remotely from one another then the shape descriptors will necessarily be different.

A natural question then becomes, can one minimize the difference function

$$D^2 = \sum_{l=0}^{L} \sum_{m=-l}^{+l} \left| a_{lm} - \bar{b}_{lm} \right|^2$$

where **a, b** are the $\mathbf{l}^{th}$ /$\mathbf{m}^{th}$ descriptors of molecule **A, B** respectively, summed over all values of **l,m**[see ref. **8** for derivation of this shape difference metric] in such a way as to give rise to a matching set of descriptors for both participants in the complex ?

If this can be achieved, then one has effectively docked one surface onto the other, and this can be checked by reconstituting the surfaces in real space in exactly the same way as for a closed surface.

In real space, the docking problem amounts to finding the right translation vector (the two surfaces must possess a common centroid) and the right rotation matrix; typically, these operations will be combined in one generalized rotation/translation matrix [13]. In fact, it is exactly this same rotation/translation matrix (invoking Parseval's theorem [10]) that is required in shape descriptor space. The advantage in doing this in shape descriptor space is that it is computationally much less expensive - the function to be minimized contains a summation instead of a double integral. The details of how this is done are to be found in ref. 10.

It can now easily be demonstrated that molecular surfaces derived from protein can be docked onto one another. One example of this is the eye-lens protein γ-crystallin, which consists of a single polypeptide of 174 aminoacid residues which folds into two (quasi-symmetrical) globular domains [14]. The regions on the surface of these two domains that are in contact (defined as those surface atoms that are with 1.4 Å or less of any atom in the other domain) form an interface with an area of 905 $Å^2$.

In a simulated docking experiment, the protein was severed into its two domains and these were separated and arbitrarily rotated with respect to one another. The `docking' transformation was then calculated, by stepping through the local Eulerian angles in order to minimize the distance function, as described earlier.

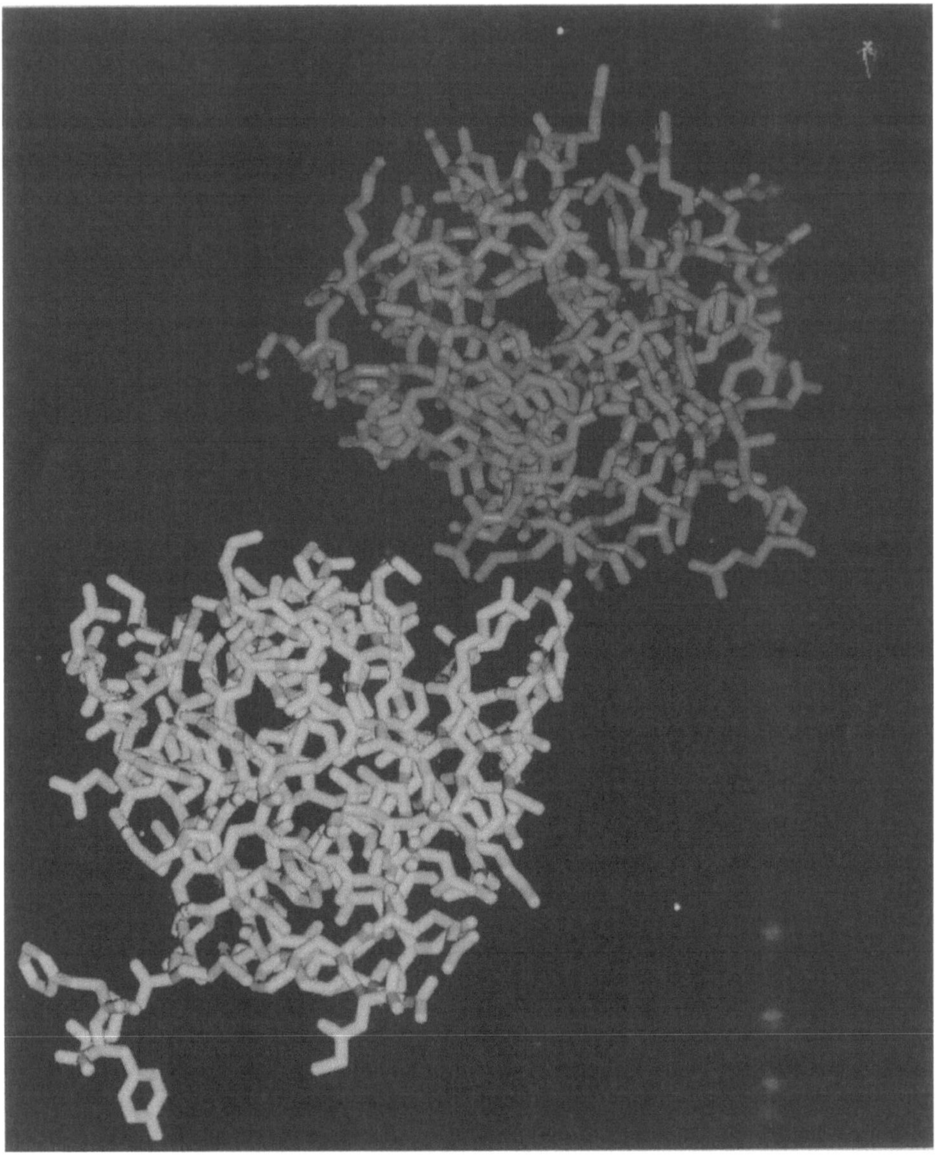

Figure 1a. Two subdomains of bovine γ-crystallin cleaved by deleting residues 83 to 86 and arbitrarily displaced from their original positions in the crystal structure.

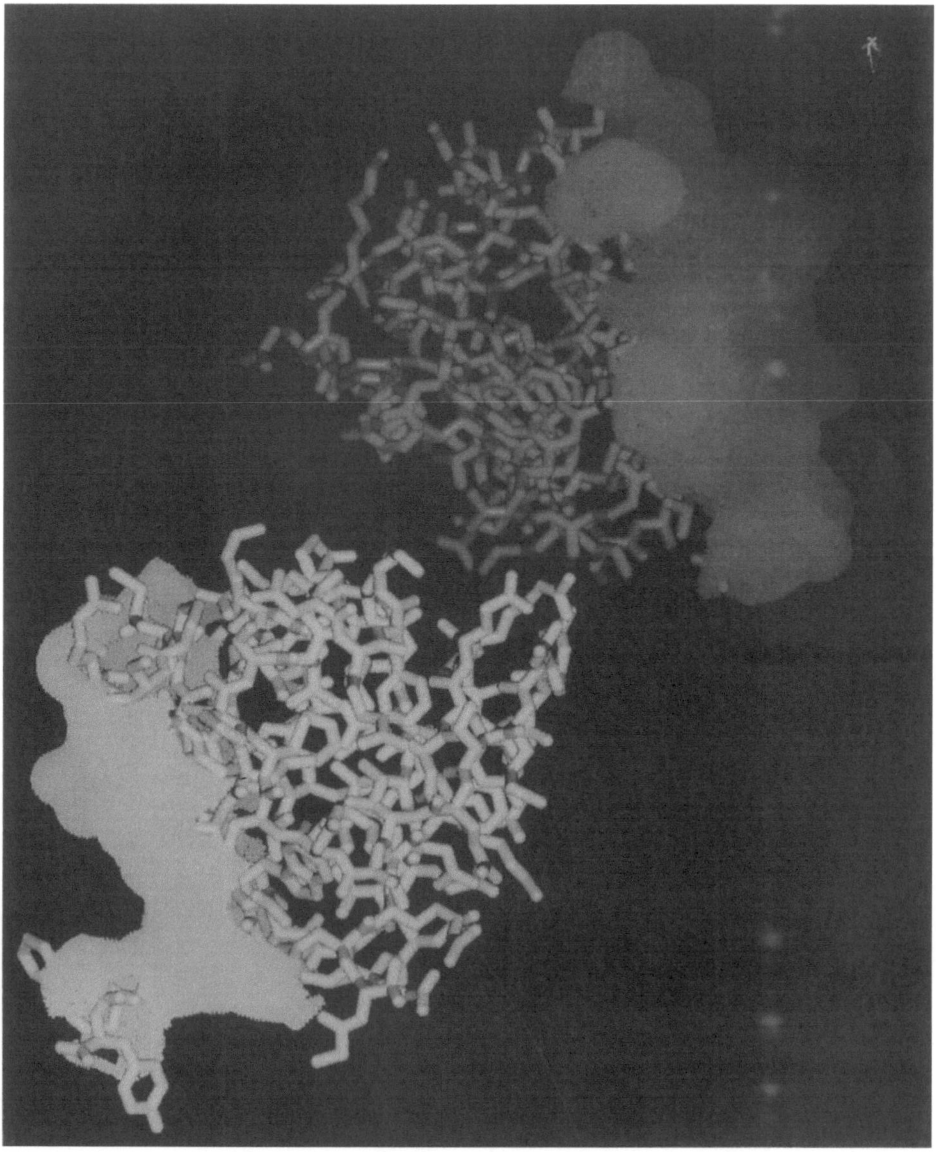

Figure 1b. The two displaced subdomains of bovine γ-crystallin with the Connolly surfaces [**11**] corresponding to their mutual binding interface added.

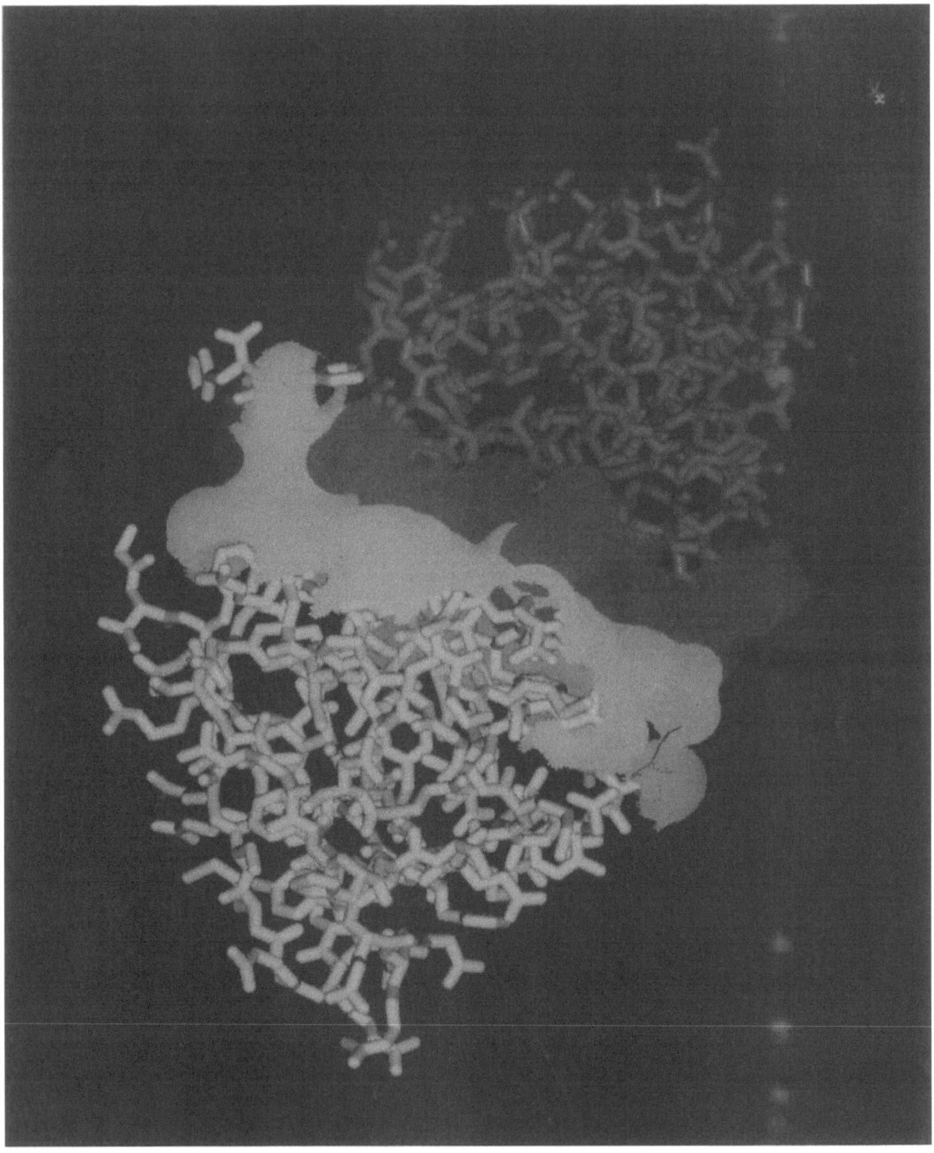

Figure 1c. The two subdomains of bovine γ–crystallin in their original location showing the intimate binding between the two halves of the protein molecule.

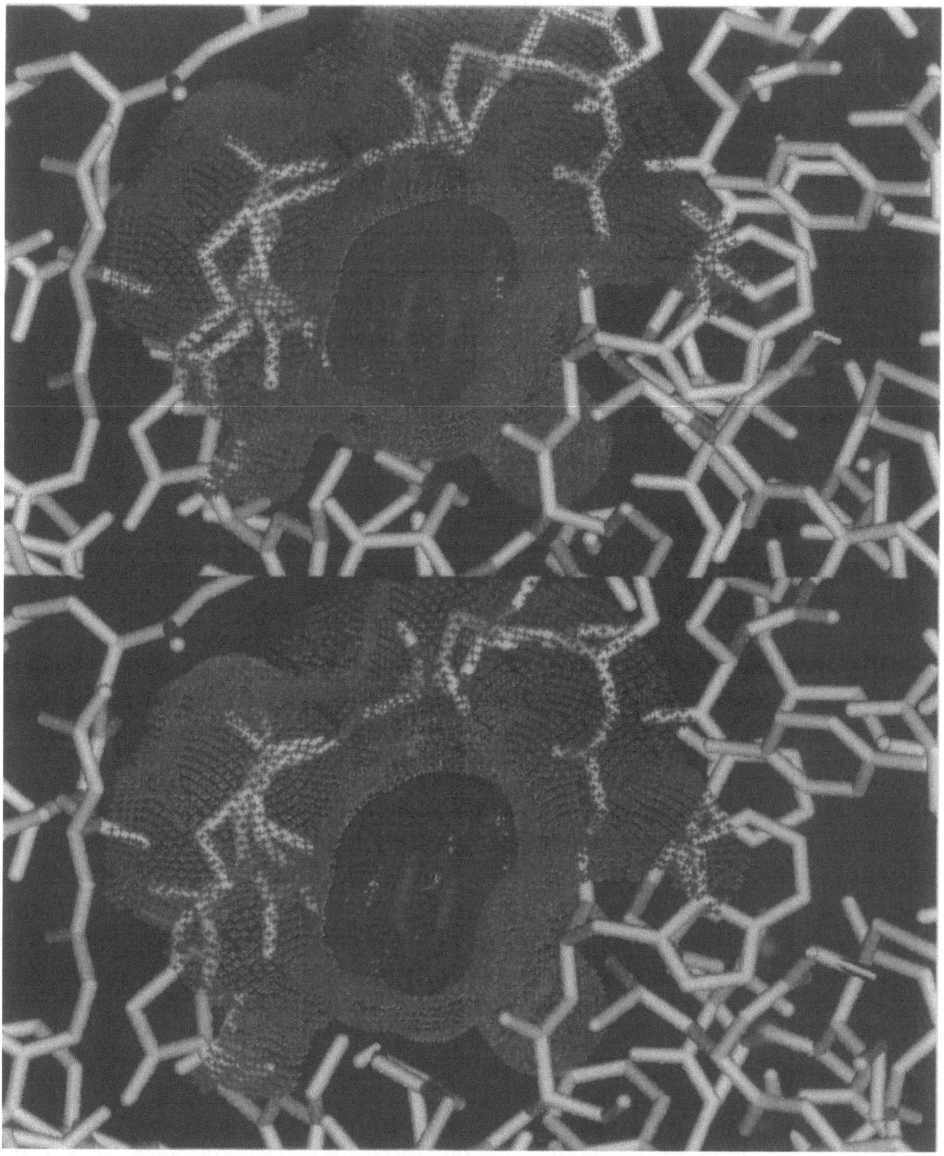

Figure 2. Stereo view of the binding of benzamidine to trypsin, showing the Connolly surfaces [**11**]. The benzamidine molecule is almost buried. Note the bulge deep in the binding pocket. This corresponds to a bound water molecule observed in the crystal structure [**15**], but not included within the Connolly envelope for the bound ligand.

The task is illustrated graphically by the series of Figures **1a, 1b, 1c**.

In Fig. **1a** the two separated halves of the molecule are shown, and it is impossible to decide by inspection how two such complex objects could be fitted together. A clue is provided in Fig. **1b** where the complementary binding surfaces have been added. But it is still not an easy matter to find out in 3 dimensions exactly how the molecules fit together. The original complex is as shown in Fig. **1c**.

When the normalization procedure (minimization of equ. **1**) was carried out, the docking was effectively accomplished [**12**].

In this work, it was observed that the best fit was obtained with rotation angles that were not the same for all of the l values. Higher-order terms contribute to the fit in a different way from the lower order terms.

In the case of protein-protein interactions, only the local interface can be considered, since taking the entire surface would include information not connected with the recognition process.

Docking of small bond ligands

In the case of small bound ligands, the protein binding pocket binds the predominant proportion of the total surface area of the ligand. In this case, it may be possible to conduct an automatic docking without any *a priori* knowledge of the binding mode of the ligand and without calculating shape descriptors for that part on the ligand in contact with the receptor binding 'pocket', (although this region on the receptor molecule, e.g. for some related ligand must be known.)

In other words, it may be possible to conduct a `docking' of a new ligand by taking a global set of shape descriptors for the entire molecule if it is sufficiently well buried. Some `*noise*' will inevitably be introduced, arising from the part of the molecule not in contact, but if the majority of the area is in contact, there is a possibility that the right docking orientation may be found. An example of this is the small synthetic trypsin inhibitor benzamidine [**15**], in which the contact area occupies 89% of the total 159 $Å^2$ surface area of the molecule (Fig. **2**).

In this case, it is very likely that the automatic docking method based on shape descriptors will work reasonably well (*this is currently under investigation*), but even in this case, there is an unmatched bulge clearly visibile in Fig. **2**, which in the complex is occupied by a water molecule [**15**]. This exemplifies only one of the pitfalls in any kind of automatic docking procedure. The existence of a water molecule in the binding pocket is not an uncommon occurrence.

A completely general docking scheme can moreover not be expected to work fully satisfactorily on the basis of global shape information. There are prospects that the required local shape information may be capable of being extracted from the global shape information, which must contain it. The methodology for doing this is being developed.

CONCLUSION

The shape descriptor method gives an accurate, quantitative description of molecular surfaces which can be used to compare shapes and establish correlations between shape and chemical/biological properties.

The binding of ligands to proteins can also be studied and 'automatic docking' of the ligands carried out if the mode of binding is known and structural information about the binding region is available. Only the local shape information will guarantee a correct docking, although for almost fully `enclosed' ligands, global shape information for the ligand may suffice. However, even in the case of global shape information, the local shape information is present, and methods for extracting this are being developed.

KEYWORDS

Molecular shape analysis, molecular recognition, molecular similarity, shape similarity, congruent surfaces, Fourrier shape descriptor.

REFERENCES

[1] Langley J.N., On nerve endings and special excitable substances in cells, Proc. Roy. Soc. B78, 170-194 (1906)

[2] Lee R.H., Rose G.D., Molecular Recognition 1: automatic identification of topographic surface features, Biopolymers, 24, 1613-1627 (1986).

[3] Åqvist J., Tapia O., Surface fractality as a guide for studying protein-protein interactions, J. Mol. Graph., 5, 30-34 (1987).

[4] Dean P.M., Callow P., Chau P.L., Molecular Recognition:blind-searching for regions of strong structural match on the surface of two dissimilar molecules, J. Mol. Graph., 6, 28-34 (1988).

[5] Colloc'h N., Mornon J.P., A new tool for qualitative and quantitative analysis of protein surfaces using B-spline and density of surface neighborhood, J. Mol. Graph., 8, 133-140 (1990).

[6] Leicester S.E., Finney J.L., Bywater R.P., Description of molecular surface shape using Fourier descriptors, J. Mol. Graph., 6, 104-108 (1988).

[7] Max N.L., Getzoff E.D., Spherical harmonic molecular surfaces, IEEE Computer Graphics and Applications, 8, 42-50 (1988).

[8] Leicester S.E., Finney J.L., Bywater R.P., A quantitative representation of molecular surface shape i) Theory and development of the method, J. Math. Chem., 16, 315-341 (1994).

[9] Leicester, S.E., Finney, J.L. and Bywater, R.P., A quantitative representation of molecular surface shape ii) Protein classification using Fourier descriptors and classical scaling, J. Math.Chem., 16, 343-365 (1994).

[10] Leicester S.E., Finney J.L., Bywater R.P., A quantitative representation of molecular surface shape iii) Correlations between molecular shape and biological activity, (in preparation).
[11] Connolly M.L., Analytical molecular surface calculation, J. Appl. Cryst., 16, 548-558 (1983).
[12] Leicester, S.E., Finney, J.L., Bywater R.P., A quantitative representation of molecular surface shape iv) Description of binding interfaces in molecular complexes and automatic docking, (in preparation).
[13] Foley J.D., Van Dam A. Fundamentals of Interactive Computer Graphics, Addison Wesley, Reading Ma. USA 194 p. 254.
[14] Chirgadze Y., Kurochkina N., Nikonov S., Molecular cartography of proteins: surface relief analysis of the calf eye-lens protein γ-crystallin, Protein Engineering, 3, 105-110 (1989).
[15] Marquart M., Walter J., Deisenhofer J., Bode W., Huber R., The geometry of the reactive Site and of the 2 peptide groups in trypsin,trypsinogen and its 3 complexes with inhibitors", Acta Crystallogr. B39 480-. (1983).

THE GENETIC CODE - PASSING FROM ONE DIMENSION TO THREE DIMENSIONS

Richard LAVERY

Laboratoire de Biochimie Théorique, CNRS URA 77, Institut de Biologie Physico-Chimique
13, rue Pierre et Marie Curie, Paris 75005, France

ABSTRACT

The genetic message contained within the base sequence of DNA is read, translated, copied and controlled by extremely subtle interactions with proteins. These proteins interact specifically with DNA not only by directly recognising the functional groups of the base pairs, but also by recognizing the local conformation and, very probably, the local flexibility and dynamics of the double helix. The present article discusses ways in which molecular modeling can contribute to understanding such interactions and to predicting the properties of the double helix as a function of its sequence.

RESUME

Le message génétique contenu dans la séquence de base de l'ADN est lu, traduit, copié et contôlé par des interactions avec des protéines qui sont remarquablement subtiles. Ces protéines interagissent spécifiquement avec l'ADN, en reconnaissant non seulement les groupements fonctionnels des bases, mais aussi la conformation locale et, très probablement, la flexibilité et la dynamique de la double hélice. Dans cet article nous discutons le rôle de la modélisation moléculaire dans la compréhension de telles interactions et la prédiction des propriétés de la double hélice en fonction de sa séquence.

1 INTRODUCTION

Today we are facing an explosion of information concerning nucleic acid sequences. This information, resulting in part from the Human Genome Project, opens many perspectives for understanding both the functioning and the malfunctioning of living organisms. However, the sequence of bases which constitutes the genetic library is not sufficient. Biological activity requires that this information first be translated from a one dimensional to a three dimensional code, that is to say, from a sequence into the local structure and dynamic behaviour of the DNA double helix (1). Only then can the code be exploited and, through a series of extremely precise protein-nucleic acid interactions, play its role in instructing cellular function.

Understanding the mechanics and the energetics which govern the fine structure of DNA as a function of its sequence will require a large structural database (2), unfortunately not currently available. Despite important progress in experimental techniques for determining macromolecular structure, both X-ray crystallography and NMR spectroscopy suffer from certain limitations in the case of oligonucleotide studies. In the case of crystallography it is often difficult to obtain high quality crystals of B-DNA and, in addition, distinguishing between intrinsic sequence effects and lattice induced deformations is not easy. In the case of NMR studies, the linear nature of the DNA molecule makes it more difficult to determine fine structure from short NOESY interactions, while COSY data only gives access to a limited number of backbone angles.

Confronted with these difficulties, we have attempted to develop modeling techniques to build a larger data base quickly and which will perhaps lead to some understanding of the underlying mechanics by which base sequence becomes 3D structure. Work carried out over the last few years enabled us to show that our approach apparently correlates well with available experimental data. It also brought to light the importance of sugar pucker in controlling DNA conformation which, in turn, led to methods for building a large structural database of DNA and the possibility of predicting the conformation of long DNA fragments.

2 METHODOLOGY

In attempting to model nucleic acid fine structure, we began with the decision to build a simulation algorithm which would take into account the stereochemistry of the nucleic acids and their commonly helical structure. We also chose to simplify our representation in order to reduce as far as possible the number of variables necessary to model a given fragment of DNA. These choices led to the Jumna (JUnction Minimisation of Nucleic Acids) algorithm (3) which combines the direct use of helicoidal coordinates for positioning individual nucleotides with respect to a common helical axis system with internal coordinates (dihedral angles and, inside the sugar rings, valence angles) for describing the internal flexibility of each nucleotide.

Junctions between successive 3'-monophosphate nucleotides are maintained using harmonic distance and valence angle constraints. In this way, we achieve a ten-fold reduction in the number of degrees of freedom representing the system compared with calculations in Cartesian coordinates. Although we necessarily lose some fine detail with our model, this is offset by improved possibilities for searching conformational space.

Jumna has another important advantage, namely, the possibility of controlling chosen features of nucleic acid conformation during the simulation. The first, and most important aspect, of this control is the possibility of introducing helical symmetry by simply grouping together sets of helically equivalent variables. This can lead to a further order of magnitude reduction in the number of variables representing the system and opens the way for treating infinite regular polymers very efficiently. The second aspect of control is the possibility of introducing a wide variety of constraints (inter-atomic distances, dihedral angles, helical parameters, sugar puckers, axis curvature, etc.). Recent extensions also allow for the automatic calculation of 1D and 2D energy maps and for inducing conformational transitions. Using Jumna we have been able to study the conformation and flexibility of both oligomers and polymers of DNA and RNA (4-9), and to investigate the transition pathways which link stable conformers (3, 10-12).

Jumna presently contains some 15,000 lines of code. Its force field FLEX (13-16) is based on Lennard-Jones terms (13) and atomic charges (14) specifically developed for treating nucleic acids. Dihedral angle barriers and valence angles have been largely parameterized on the basis of quantum chemical calculations. Water and counter ion effects are currently reduced to a simple electrostatic screening using a sigmoidal distance dependent dielectric function (3,17) and reduced net charges on each phosphate group (-0.5e). In order to analyse and compare nucleic acid structures we use the Curves algorithm (18,19) which calculates both the ideal axis representing a given nucleic acid fragment and a full set of helicoidal parameters.

3 RESULTS

The modeling techniques described above were put to use in a systematic attempt to understand the origins of base sequence effects on DNA. Our first studies were limited to the ten unique dinucleotide steps, which were studied with the help of six DNA polymers with regular repeating base sequences (4):

$(AA)_n$	AA	TT*		
$(GG)_n$	GG	CC*		
$(CG)_n$	CG	GC		
$(TA)_n$	TA	AT		
$(CA)_n$	GA	AC	TG*	GT*
$(GA)_n$	GA	AG	TC*	CT*

(Note that although the four nucleic acid bases can be combined in 16 different ways to make a two-base sequence, there are in fact only ten unique dinucleotide junctions. Six combinations, marked above with asterisks, are simply the complementary strands of the dinucleotide occurring on the same line.)

These polymers were subjected to helical symmetry constraints and energy was minimised to find the most stable conformations of each sequence. End effects were avoided by minimising the energy per repeating unit of each polymer taking into account seven nucleotides on either side of the central dinucleotide unit. The results obtained can therefore be considered to apply to infinite regular polymers. The conformational space belonging to the B-DNA family of structures was studied as thoroughly as possible by using many starting points for each minimisation and by deforming and then relaxing the conformations obtained to overcome conformational energy barriers. The results of this study led to the finding that each base sequence was associated with a number of stable sub-states having similar energies (Table 1).

Table 1. Energy per dinucleotide repeat unit for the stable sub-states of the regular DNA polymers studied.

Sequence	Energies (Kcal/mole)					
$(AA)_n$	-82.5*	-82.4*	-81.9*			
$(GG)_n$	-118.1	-115.2	-115.1*	-114.2*		
$(CG)_n$	-119.4	-118.7	-118.0	-114.0#		
$(TA)_n$	-81.6	-80.8	-80.2	-80.0#		
$(GA)_n$	-99.5	-98.4#	-98.3	-97.6	-97.0	-96.7#
$(CA)_n$	-100.9	-100.8	-100.6	-100.4	-100.3	-100.2
	-99.6	-99.6	-99.2	-97.2#	-97.1#	-95.2#
	-95.0#					

Note: Sub-states having purines with O1'-endo sugars are indicated by the symbol #. Sub-states of homopolymers showing dinucleotide symmetry are indicated by the symbol *.

Although the mean values obtained for both backbone and helicoidal parameters were close to experimental values, it was found that the various sub-states of each sequence often had very different conformations (4, 20). Further investigation of the sub-state conformations showed that the set of sub-states associated with each base sequence could be distinguished on the basis of their sugar puckers alone (Table 2). The sugars were found to lie within three pseudorotational families which we have named S (C2'-endo puckers with low amplitude), X (C2'-endo puckers with high amplitude) and E (O1'-endo puckers, associated commonly, although not exclusively, with pyrimidine nucleotides). These sugar families were found to be correlated with a number of other structural parameters and thus to have a major role in determining the local conformation of the double helix. These findings were subsequently confirmed by modifying Jumna to automatically calculate energy maps as a function of sugar pucker.

All the sub-state minima previously found could be located on these maps, which, in addition, allowed us to visualise the transition pathways and the energy barriers separating the different sub-states. In general it was found that sub-states containing only S and X sugars were separated by very small energy barriers below 1 Kcal/mole, while transitions from these sub-states to those containing E sugars required passing barriers of roughly 2 Kcal/mole. This study also allowed us to demonstrate the direct coupling of sugar pucker to a variety of other parameters including both helicoidal variables and backbone dihedrals (20).

Table 2. Sub-states classified according to their sugar puckering (S: C2'-endo, low amplitude, X: C2'-endo, high amplitude, E: O1'-endo). The sugars in each strand are given in the 5'→3' sense and follow the order of the bases given in the first column of the table.

Sequence	Sugar puckers					
$(AA)_n$	SX:SS*	SX:ES*	SX:SE*			
$(GG)_n$	XX:XX	SS:XX	XS:SX*	SX:ES*		
$(CG)_n$	SX:SX	XS:XS	ES:ES	SE:SE#		
$(TA)_n$	SX:SX	XS:XS	ES:ES	XE:XE#		
$(GA)_n$	XS:SX	ES:SX#	SX:ES	XS:SE	SX:SE	ES:SE#
$(CA)_n$	ES:XS	XS:SX	XS:XS	ES:SX	SX:XS	ES:ES
	XS:ES	SX:SX	SX:ES	SE:XS#	SE:XX#	SE:SE#
	SE:ES#					

Note: Sub-states having purines with O1'-endo sugars are indicated by the symbol #. Sub-states of homopolymers showing dinucleotide symmetry are indicated by the symbol *.

Although these results led to some interesting data on the nature of base sequence effects and, notably, on the role of sugar pucker, it was also clear that explaining available experimental data would require going beyond the nearest neighbour dinucleotide model (2).

If we are concerned mainly with internucleotide parameters such as rise, twist and wedge angles, then the next step beyond the dinucleotides involves studying tetranucleotide sequences. This implies the passage to a next nearest neighbour model where each unique dinucleotide step will now be studied in the context of all possible 5' and 3' neighbouring base pair steps. This considerably complicates the study to be undertaken, since a rapid calculation shows that there are a total of 136 unique tetranucleotide sequences.

Although these sequences can again be grouped into a smaller number of regular polymers. there are nevertheless 39 polymers to be studied compared to the six necessary for the dinucleotides. In addition, each polymer must be treated using tetranucleotide rather than dinucleotide symmetry constraints leading to eight unique sugars per repeating unit.

The size of this computational problem necessitated the development of automated procedures for finding sub-state energy minima. The procedure we developed is based on the fact that all sub-states found for each dinucleotide sequence could be uniquely labelled by their sugar puckers. This led the idea of a combinatorial search based on the puckers of each sugar in the repeating unit as illustrated in Figure 1. For each step of this search the symmetry unique sugars were first constrained to a given set of puckers and an energy minimisation was performed. The constraints were then released and a second minimisation was carried out.

Figure 1. Combinatorial searching on the basis of the sugar puckering of each nucleotide in the repeat unit of a regular DNA polymer.

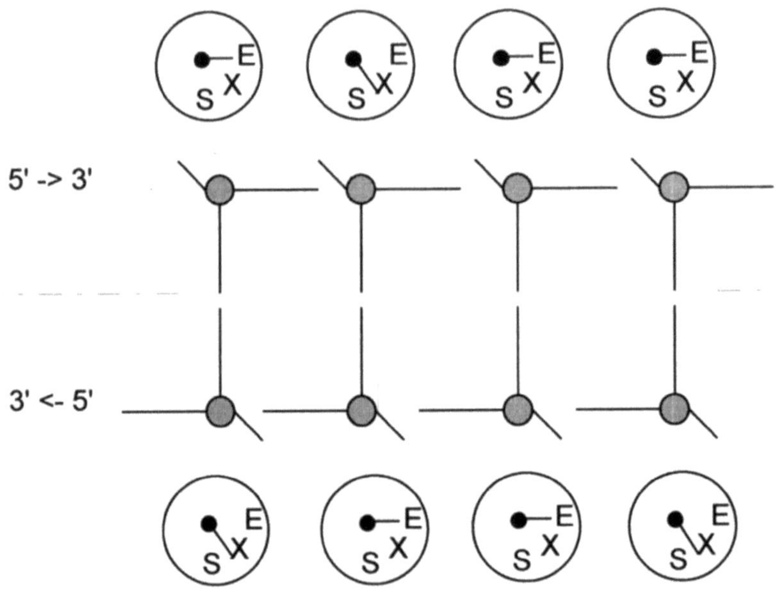

The results of this study showed that there are on average 20 sub-states per polymer, individual cases varying from a minimum of 5 to a maximum of 60. In addition, since many sequences possessing mononucleotide, dinucleotide or inversion symmetry actually have stable conformations which do not express this symmetry (8, 20, 23), the total number of sub-states is increased by multiple copies with identical energies and related by the symmetry operations applicable to the polymer in question. Note also that although only S, X and E sugar puckers were used as starting points in the combinatorial search, certain sub-states are generated with C3'-endo sugars, usually characteristic of A-DNA, while other sub-states contain unusual backbone geometries involving the ε/ζ or α/γ dihedral angles.

From these results it is possible not only to understand more clearly the extent and the nature of base sequence effects, but also to attempt to construct longer sequences of DNA by tetranucleotide superposition. If we assume that next nearest neighbour interactions will generally suffice to describe sequence effects, then the tetranucleotide database we have constructed already contains the information needed to assemble a DNA fragment with any chosen sequence.

The technique we have developed for this construction is illustrated schematically in Figure 2 for the sequence CATGACGTCATG containing the Cre site (24) whose structure has already been obtained using NMR data (7). Firstly, we decompose the sequence into its constituent tetranucleotides which consequently overlap by trinucleotide sequences. Next we search the database for tetranucleotide sub-states which have the same structure as their neighbours for the six overlapping nucleotides. As indicated in Figure 2, each overlap is tested using an angular RMS value which takes into account the phosphodiester backbone angles (α, β, γ, ε, ζ), the sugar conformation (phase, amplitude) and the glycosidic angle (χ) of each nucleotide. Excluding backbone linkages outside the trinucleotide overlapping fragment leads to a total of 38 parameters which contribute to the RMS value calculated.

It might be suspected that since neighbouring tetranucleotides within an irregular sequence generally come from different initial polymer studies there would be little chance of obtaining good superposition. However, the tests we have carried out so far suggest that this is not the case and successful superpositions can generally be made with a remarkable quality of around 3°→4° RMS.

Figure 2. Superposition technique for building long sequences from the tetranucleotide database.

(a) Tetranucleotides composing the Cre sequence CATGACGTCATG

C	A	T	G	A	C	G	T	C	A	T	G
C	A	T	G								
	A	T	G	A							
		T	G	A	C						
			G	A	C	G					
				A	C	G	T				
					C	G	T	C			
						G	T	C	A		
							T	C	A	T	
								C	A	T	G

(b) Variables taken into account during the superposition tests (indicated in the schematic diagram by bold lines). After successful superposition the two nucleotide indicated by the black circles are added to the polymer under construction.

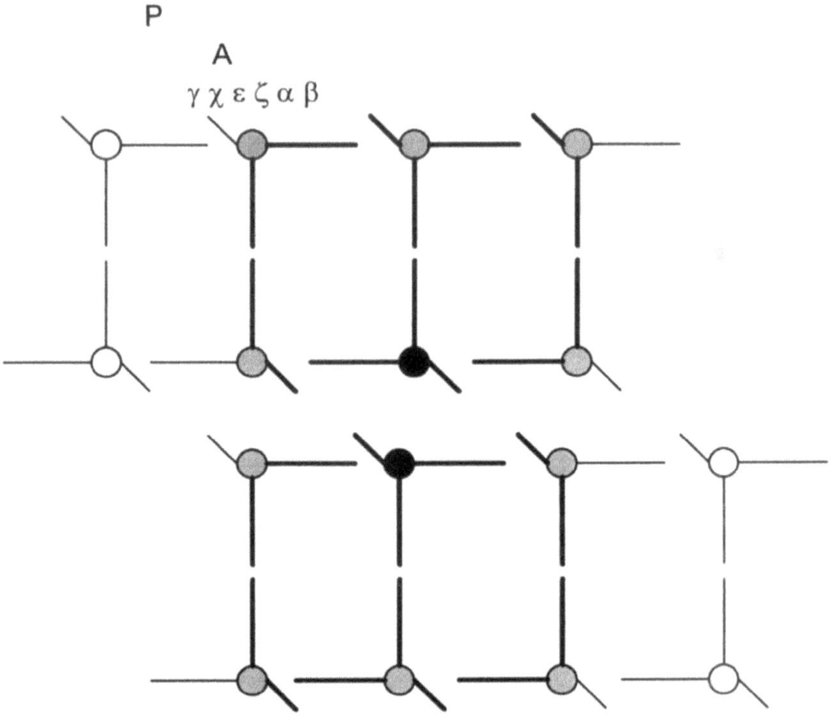

Our second finding concerns the number of conformations which can be constructed for a given sequence. In the case of the Cre sequence, the number of sub-states in the database for each of its constituent tetranucleotides enables us to calculate that there are roughly $3x10^{12}$ possible structural combinations. However, once superposition is taken into account a very different picture emerges. As seen in Table 4, fitting together tetranucleotides in an irregular sequence leads to a dramatic decrease in structural multiplicity and, in general, the final DNA fragment is compatible with only a small number of global conformations. For Cre, the central tetranucleotide ACGT begins with 45 stable sub-states. Adding the two neighoouring tetranucleotides (GACG on the 5' side and CGTC on the 3' side) and searching for fits of better than 4° RMS eliminates all but 11 of the ACGT sub-states and simultaneously half of the 22 sub-states of the adjoining tetranucleotides. If we now continue the construction by adding further tetranucleotides at either end of the central fragment, the number of acceptable ACGT sub-states is reduced successively to eleven, then four possibilities and, finally, to a single sub-state. At the end of the construction all of the central part of the oligomer has a unique structure and only its ends maintain a number of possible conformations. It should also be noted that although we only use tetranucleotide fragments for this construction, this does not exclude the appearance of sequence effects well beyond the next nearest neighbour range. In the case discussed, the conformation of the central nucleotides of the Cre sequence is finally chosen by fixing the base sequence five nucleotides away from the centre of the oligomer.

Table 3. Successive sub-state elimination during the construction of the Cre sequence CATGACGTCATG (the subscript i indicates that two sub-states n and ni are related by inversion symmetry).

								A		**C**		**G**		**T**								
											1 → 45											
						G		**A**		**C**		**G**		**T**		**C**						
									1		1		1									
									2		9		2									
									5		9i		5									
									6		10		6									
									7		16		7									
									8		16i		8									
									11		19		11									
									12		25		12									
									13		25i		13									
									16		27		16									
									17		27i		17									
				T		**G**		**A**		**C**		**G**		**T**		**C**		**A**				
							8		2		1		2		8							
							12		5		10		5		12							
							17		11		16		11		17							
											16i											
		A		**T**		**G**		**A**		**C**		**G**		**T**		**C**		**A**		**T**		
					3		8		2		1		2		8		3					
					6												6					
C		**A**		**T**		**G**		**A**		**C**		**G**		**T**		**C**		**A**		**T**		**G**
			1		3		8		2		1		2		8		3		1i			
			3		6												6		3			
			7i																7			
			16i																16			
			21i																21			
			30																30i			

Having built a conformation for our oligomer, we must now determine whether this conformation corresponds to that obtained by direct minimisation of the fragment. The answer is contained in Table 5 which shows that, with the exception of the terminal nucleotides, the database constructed fragment coincides almost perfectly with one of the two conformations found for the Cre dodecamer in our earlier study (7). (Note that this comparison was made with an arbitrary choice for the terminal nucleotides of the fragment which, as shown in Table 4, have a number of possible conformations. It should be remarked that one should not in any case expect a perfect agreement at the ends of an oligomer since the database conformations all belong to infinite regular polymeric sequences. Consequently, the explicit end-effects, present in the minimised oligomer, certainly cannot be reproduced).

Table 4. RMS fit between the structure of the Cre dodecamer CATGACGTCATG obtained by minimisation (conformation Ib in ref. 5) and that obtained by construction from the tetranucleotide database (values are in degrees, 8 angular variables are compared for each nucleotide).

Res	γ	ε	ζ	α	β	χ	Pha	Amp	Rms
C1	---	4.1	39.8	2.4	7.7	32.5	77.3	6.0	35.3
A2	3.3	0.3	0.1	1.5	2.3	6.9	1.5	0.3	2.93
T3	1.3	0.7	0.2	1.0	1.2	2.8	0.1	0.9	1.29
G4	1.6	0.9	2.6	1.2	1.1	5.8	8.2	1.4	3.82
A5	1.4	1.7	3.4	1.1	0.7	4.4	0.7	0.1	2.16
C6	0.3	0.7	2.8	1.3	0.7	1.4	2.5	0.1	1.54
G7	0.2	0.9	1.8	0.9	3.7	3.5	3.5	1.0	2.35
T8	1.4	0.1	7.4	2.0	0.2	1.9	0.3	1.5	2.88
C9	1.5	0.1	2.0	0.1	2.4	0.9	6.8	2.1	2.81
A10	0.1	0.0	1.7	0.0	2.7	3.5	2.3	0.7	1.87
T11	1.0	0.2	0.6	0.9	6.3	1.0	7.4	1.3	3.52
G12	2.5	---	---	---	---	8.8	12.0	5.8	8.08
G13	2.6	---	---	---	---	8.7	12.0	5.7	8.08
T14	1.0	0.2	0.7	0.8	6.2	1.0	7.3	1.3	3.49
A15	0.2	0.0	1.7	0.1	2.6	3.6	2.2	0.7	1.87
C16	1.5	0.1	2.0	0.1	2.5	0.8	6.9	2.1	2.84
T17	1.4	0.1	7.4	2.1	0.1	2.0	0.4	1.5	2.90
G18	0.2	0.9	1.8	0.9	3.8	3.5	3.6	1.1	2.38
C19	0.4	0.7	2.8	1.2	0.6	1.4	2.5	0.1	1.53
A20	1.4	1.7	3.5	1.1	0.6	4.3	0.7	0.1	2.17
G21	1.5	0.9	2.7	1.3	1.1	5.8	8.3	1.4	3.83
T22	1.3	0.7	0.2	1.0	1.2	2.8	0.1	0.9	1.28
A23	3.2	0.2	0.1	1.4	2.4	7.0	1.5	0.3	2.94
C24	---	4.0	39.8	2.3	7.7	32.5	77.3	6.0	35.3

CONCLUSION

Our studies of base sequence effects on DNA conformation suggest that all repetitive sequences are associated with a number of stable sub-states which are close in energy, but can vary widely in conformation. The passage from dinucleotide to tetranucleotide symmetry considerably increases the number of sub-states associated with each sequence. However, it appears that building an irregular sequence by tetranucleotide superposition results in a dramatic reduction in structural multiplicity.

These results suggest that it should indeed be possible to construct longer fragments of DNA using the tetranucleotide database. This approach will enable us to overcome the computational barriers which would normally hinder the study of long fragments. One important application in view is certainly the problem of sequence dependent DNA curvature. In this area it should be stressed that all present prediction techniques are base on a nearest neighbour model and thus only distinguish between the ten unique dinucleotide junctions (25-27). In contrast, our database construction is based on a next nearest neighbour model and can distinguish between the 136 unique tetranucleotides sequences.

ACKNOWLEDGMENT

We wish to thank the Association for International Cancer Research (St. Andrews University, UK) for their generous support of this research project and the French supercomputer center IDRIS for their allocation of computer resources.

KEYWORDS

Genetic code, helical symmetry, sub-state, base sequence, nucleic acids, molecular modeling

REFERENCES

[1] Dickerson R.E.in Unusual DNA Structures,. Wells R.D and Harvey S.C. (eds) Springer-Verlag (1988).
[2] Yanagi K., Privé G.G., Dickerson R.E., J. Mol. Biol. 217, 201 (1991).
[3] Lavery R. in Structure and expression, Vol.3, DNA bending and curvature, Olson W.K., Sarma R.H., Sarma M.H., Sundaralingam M. (eds) Adenine Press, New York, p.191 (1988).
[4] Poncin M., Hartmann B., Lavery R., J. Mol. Biol. 226, 775 (1992).
[5] Poncin M., Piazzola D., Lavery R., Biopolymers 32, 1077 (1992).
[6] Zakrzewska K., J. Biomol. Struct. Dynam. 9, 681 (1992).
[7] Mauffret O., Hartmann B., Convert O., Lavery R. , Fermandjian S., J. Mol. Biol. 227, 852 (1992).
[8] Zakrzewska K., Poltev V.I., Oguey C., R. Lavery, J. Mol. Struct. (Theochem) 286, 219 (1993).
[9] Ouali M., Letellier R., Adnet F., Liquier J., Sun J.S., Lavery R., E. Taillandier, Biochemistry 32, 2098 (1993).
[10] Hartmann B., Piazzola D., Lavery R., Nucleic Acids Res., 21, 561 (1993).
[11] Briki F., Ramstein J., Lavery R., Genest D., J. Am. Chem. Soc. 113, 2490 (1991).

[12] Sanghani S.R., Lavery R., Nucleic Acids Res. 22, 1444 (1994).
[13] Zhurkin V.B., Poltev V.I. , Florent'ev V.L., Mol. Biol. 14, 1116 (1980).
[14] Lavery R., Zakrzewska K., Pullman A., J. Comp. Chem. 5, 363 (1984).
[15] Lavery R., Sklenar H., Zakrzewska K., Pullman B., J. Biomol. Struct. Dynam. 3, 989 (1986).
[16] Lavery R., Parker I., Kendrick J., J. Biomol. Struct. Dynam. 4, 443 (1986).
[17] Hingerty B., Richie R.H., Ferrel T.L., Turner T.E., Biopolymers 24, 427 (1985).
[18] Lavery R., Sklenar H., J. Biomol. Struct. Dynam. 6, 655 (1989).
[19] Lavery R., Sklenar H., in Structure and methods, Vol.2, DNA Protein complexes and proteins. Sarma R.H., Sarma M.H.(eds), Adenine Press, p.215 (1990).
[20] Lavery R., Hartmann B., Biophys. Chem. 50, 33 (1994).
[21] Lipanov A., Kopka M.L., Kaczor-Grzeskowiak M., Quintana J., Dickerson R.E., Biochemistry 32, 1373 (1993).
[22] Arnott S., Chandrasekaran R., Birdsall D.L., Leslie A.G.W., Ratliff R.L., Nature 283, 743 (1980) and coordinates communicated to our laboratory by Arnott S.
[23] Lavery R.in Advances in Computational Biology, Vol. 1 Ed. H.O. Villar, JAI Press, onnecticut, p69 (1994).
[24] Montminy M.R., Sevarino K.A., Wagner J.A., Mandel G. , Goodman R.H., Proc. Natl. Acad. Sci. (USA) 83, 6682 (1988).
[25] Bolshoy A., McNamara P., Harrington R.E., Trifonov E.N., Proc. Natl. Acad. Sci. (USA) 88, 2312 (1991).
[26] Cacchione S., De Santis P., Foti D., Palleschi A., Savino M., Biochemistry 28, 8076 (1989).
[27] Tan R.K.-Z., Harvey S.C., J. Biomol. Struct. Dynam. 5, 497 (1987).

A NEW METHOD FOR UNBIASED COMPARISON OF PROTEIN STRUCTURES

Ramaswamy NILAKANTAN and Rengachari VENKATARAGHAVAN

Lederle Laboratories, Pearl River, NY 10965 USA

ABSTRACT

While there exist several excellent algorithms for protein sequence comparison, the comparison of protein structures is more difficult. This paper presents a method for a sequence-independent, unbiased comparison of protein structures. The method treats each protein as a collection of α-carbons without any regard to sequence or chain-connectivity. Heuristics are used to select a series of subsets of atoms in each of the protein structures. These subsets of atoms are then compared using a subgraph-isomorphism search technique. Partial matches thus obtained are then used to superpose one protein onto the other. A simple scoring function is then used to determine the validity of the match. This method has been applied to a set of about 100 diverse proteins. Several proteins known to have similar structures were identified by the method. A few protein-pairs not widely recognized to have structural similarities were also identified.

RESUME

Alors qu'il existe plusieurs excellents algorithmes pour la comparaison séquentielle des protéines, la comparaison de la structure des protéines est plus difficile. On présente ici une méthode indépendante des séquences pour une comparaison impartiale de la structure des protéines. La méthode considère chaque protéine comme un ensemble d'α-carbones, sans aucune considération séquentielle ou de connectivité de chaine. Des procédures heuristiques sont employées pour selectionner des séries de sous-groupes d'atomes dans les structures de protéines. Ces sous-groupes d'atomes sont ensuite comparés entre eux par une technique de recherche d'isomorphisme des sous-graphes. Des concordances partielles ainsi obtenues sont ensuite utilisées pour superposer une protéine sur une autre. Une simple fonction de pondération est ensuite utilisée pour apprécier la valeur de la concordance. Cette méthode a été appliquée à une population d'une centaine de protéines variées. Plusieurs protéines ayant des structures similaires connues ont été identifiées par cette méthode. Quelques paires de protéines, en général sans reconnaissance universelle de leurs similarités structurelles, ont aussi été identifiées.

1 INTRODUCTION

Proteins are being crystallized and solved at an ever-increasing rate. As the number of known proteins becomes larger, it becomes increasingly difficult to understand the structural relationships between them. When a new protein is solved, it is customary to examine its relationships with other proteins of the same class if they are available. However, it is difficult to compare a new protein with no sequence similarity to previously solved proteins with all the known proteins. There are few people who can memorize the 3-dimensional structures of proteins and discover similarities quickly. Moreover, as the number of proteins becomes larger, even they will find it difficult to hold the structural information in their memory.

Thus it is of great interest to have a fast method for structure-based alignment unbiased by sequence. The present work is one attempt at solving the problem of finding shared substructures in unrelated proteins.

Recently, several methods have appeared in the literature for carrying out comparisons of tertiary structures of proteins. Some of these are described below.

1. Shoichet and Kuntz (1991) have published a docking method for proteins which characterizes the surface features such as clefts and bumps by sets of spheres. Two proteins are matched by matching their spheres using combinatorial algorithms. This method obviously focusses only on surface features. Hence it is good for matching closely related proteins with similar surface features, but will fail in the case of unrelated proteins.

2. Jiang and Kim (1991) have described a method for docking proteins by dividing them into cubes. A simple translation-rotation search is then used to achieve the best possible docking. This is a very time-intensive approach suitable only for small numbers of comparisons.

3. Mitchell et al. (1989) describe a method by which proteins are reduced to their secondary structure elements such as helices and strands which in turn are represented by a labeled graph. A subgraph isomorphism algorithm is then used to find commonalities between the two proteins. This method obviously works only on regions of proteins with well-defined secondary structural elements. Furthermore, the method will work only when two equivalent secondary structure elements are of the exact (or nearly exact) length.

4. Vriend and Sander (1991) have published a method which aligns small segments of two proteins and then clusters these alignments into larger groups to find possible

dockings between the proteins. The problem with this method is its dependence on the initial alignment of small fragments. This means that unless small contiguous segments of the polypeptide chain are similar it is difficult to identify clusters and possible superpositions.

5. Go and coworkers (1992) have a method similar to (4) except that they do not use sequence similarity. They compare fragements by RMS fit and then examine all dockings between these fragments.

6. Fisher et al. (1993) describe a novel method based on principles of machine vision to detect similarities between protein substructures. They use a geometric hashing method to generate a set of 'seed matches' which are then clustered to identify larger regions of similarity. This method is rapid and is independent of sequence or even the linear order of residues in the polypeptide chain.

7. May and Johnson (1994) have used a genetic algorithm for automatic superposition of proteins. A string of bits encodes the translation and rotation of one protein with respect to the other. Mutations and 'crossovers' are carried out and a scoring function is used to evaluate the superposition. The method however, is not geared for large numbers of pairwise comparisons.

2 METHOD

The present approach is to consider the two proteins simply as collections of alpha-carbons and see whether there is a possible common substructure between them. In other words, if we represent each protein by a graph where each node represents an alpha-carbon and each edge a distance, the present problem is to find isomorphic subgraphs.

The problem with this approach, as has been pointed out by several authors, is that the subgraph isomorphism problem is NP-complete and even for fairly small proteins, the computation becomes overwhelming. In other words, a complete subgraph isomorphism search is a brute force approach; our approach is to choose subsets of atoms (nodes in the graph) and restrict the subgraph isomorphism searches to these subsets. Specifically, each protein is divided into concentric shells of atoms around its centroid. The shell thickness is determined by the size of the smaller of the two proteins being compared. We use the following heuristic formula to determine the thickness.

$$\text{Thickness} = a * \text{sqrt}(\min(na,nb)) + b$$

where 'a' and 'b' are parameters derived on the basis of an examination of several different pairs of proteins, and 'na' and 'nb' are the numbers of a-carbons in the two proteins.

We use another heuristic formula to determine the tolerance to be used in matching pairs of distances in the two proteins.

$$\text{Tolerance} = \min(na,nb) * c + d$$

where as before, 'c' and 'd' are parameters derived heuristically.

In the studies discussed here, the values of the parameters we used are as follows:

$a = -0.3571683$

$b = 10.48829$

$c = -0.005490196$

$d = 2.389803916$

The shells of atoms are then compared by constructing a docking graph and detecting cliques in it (Kuhl, Crippen & Friesen, 1984).

The result of the matching procedure is typically one or more matched pairs of atoms from the two proteins being compared. For each such match, one protein is docked onto the other and a score is calculated. The score is the number of alpha carbons of the smaller protein within 3 A of an alpha carbon of the other protein divided by the number of alphà carbons in the smaller protein. Thus we obtain a score on a scale of 0-1.

Low scores of less than 0.4 usually do not represent valid dockings. Scores of 0.6 or more are almost always good matches.

3 Results and Discussion

To test the present method, we carried out a study of a set of about 100 functionally diverse proteins (Boberg et al. 1992) taken from the Brookhaven Protein Data Bank. The aim of this study was to compare all pairs of proteins in this set and look for non-obvious structural similarities. Since this involves around 5000 pairwise comparisons, we used certain heuristics to trim the size of the problem as discussed below.

Pre-filter to determine which pairs of proteins to compare: The filter we have devised is based on the end-to-end lengths of small fragments of the polypeptide chain. Specifically, we created population-distributions of the end-to-end lengths of peptides of length 3 thru 8. For each pair of proteins in the set, correlation-coefficients were computed for the corresponding pairs of end-to-end length distributions of the two proteins. If even one correlation coefficient was less than 0.2, that pair of proteins was deemed too structurally different from each other to require detailed comparison. By this means, we trimmed down the set of proteins to be compared by the subgraph-isomorphism method by about a factor of 3.

We carried out pairwise comparisons on the pairs of proteins that survived the pre-filter. Several pairs of proteins that are closely related by sequence and structure were found by the method. There were also some pairs that are not related closely by sequence but have interesting structural resemblances.

Some of these pairs are discussed below.

(i) *β-trypsin-BPTI complex with α-chymotrypsin (.6cha-2ptc)*: The program was able to detect the structural similarity of the trypsin and chymotrypsin molecules despite the presence of the bound BPTI molecule. (Note that the program knows nothing about sequence or even the chain-connectivity) Thus, even though the two structures were not closely similar in overall shape, the program succeeded in finding the similarity.

(ii) *Plastocyanin-Pseudoazurin (1pcy-2paz)*: This is a very interesting case (Vriend and Sander (1992) have previously pointed out the similarity between plastocyanin and azurin) of two electron-transport proteins which are structurally extremely close but have only about 25% sequence identity. Plastocyanin has 99 residues while pseudoazurin has 124. Their structures match to a very high degree.

(iii) HIV-protease and RSV-protease: Both these proteins are viral proteases and have about 32% sequence identity. HIV-protease has 99 residues while RSV-protease is slightly larger with 124 residues. The program successfully finds the structural similarity between this pair of proteins.

(iv) *Myoglobin and 434-repressor:*(**1R69-1MBD**): This is an interesting case of shared substructures among proteins not related by function or sequence. Figure 1 shows the 434-repressor molecule (69 residues) along with the shared region of the much larger myoglobin molecule (153 residues). Myoglobin is an oxygen transport protein containing 7 helices connected by loops. the 434-repressor protein is a regulatory protein with 5 short helices connected by loops. Our matching procedure revealed that a significant 3-helix core is shared by the two proteins. While this core is only a small part of the larger myoglobin molecule, it involves more than 60% of the smaller 434-repressor. This striking similarity has been pointed out also by Subbiah et al. (1992).

(v) *Ferredoxin and Ubiquitin:* (**3FXC-1UBQ**) : These two proteins share no significant sequence similarity, nor are they functionally related. Ferredoxin is an electron transport protein while ubiquitin is a protein found in chromosomes. However, they have a similar 3-dimensional structure (Figure 2) as has recently been pointed out by Vriend and Sander (1992). These two proteins also do not have a common function.

(vi) *Barley chymotrypsin inhibitor and chloramphenicol acetyltransferase:* (**2CI2-1CLA**).: This is a very interesting case not reported before. The chymotrypsin inhibitor is a small protein of 83 residues while chloramphenicol acetyltransferase is a huge trimeric protein each subunit of which has 213 residues. The latter protein is found in chloramphenicol-resistant strains of bacteria and renders the chloramphenicol molecule ineffective. The two proteins share a significant common motif (Figure 3) consisting of a few strands of sheet and a small stretch of alpha-helix.

(vii) Elongation factor and Ubiquitin: (**1ETU-1UBQ**) : This pair of proteins again shares no functional or sequence similarity. However, they have interesting regions of structural similarity as shown in Figure 4. This has also not been pointed out by other groups so far.

1R69-1MBD

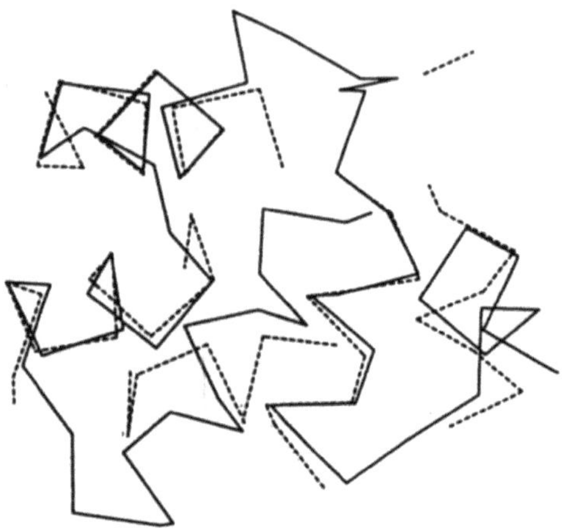

3FXC-1UBQ

Figures 1 and 2

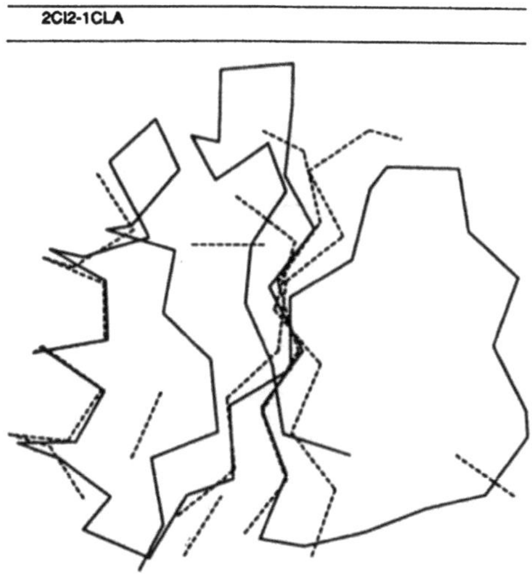

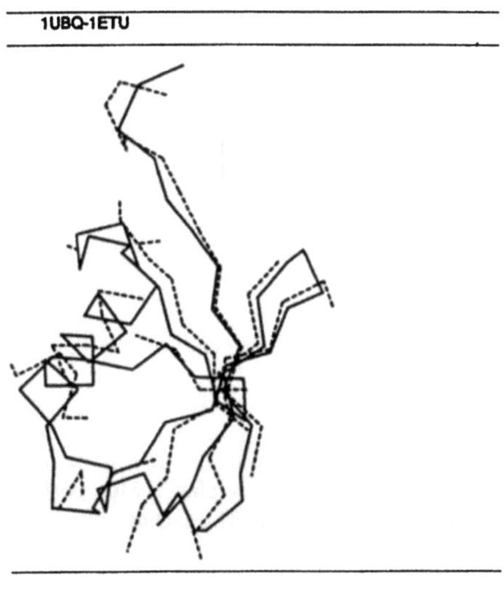

Figures 3 and 4

Advantages of this approach

1. Only coordinates, no sequence, no secondary structure information required.

2. The method is independent of chain sense; in other words, an alpha-carbon chain could in principle be matched with one of the opposite sense.

3. The method is reasonably fast; for pairs of proteins the method typically takes from a few seconds to a few minutes on the Cray YMP using a single processor.

Disadvantage of this approach

We might miss matches. The choice of atom subsets is based on a set of heuristics, which work in most cases but might conceivably fail.

Conclusion

Tertiary structural motifs consisting of disjointed sets of secondary structural elements are often found to be shared by unrelated proteins. These motifs may simply be structural scaffolds which have been conserved in evolution; on the other hand, they might simply be the consequence of the packing of periodic structures such as helices and sheets.

Keywords

Protein structures, common substructure, α-carbon method, subgraph isomorphism, shell thickness, matching pairs, docking graph, atom pairs matching.

References

[1] Shoichet B.K , Kuntz I.D., Protein docking and complementarity, J. Mol. Biol. 221, 327-346 (1991)

[2] Jiang F., Kim S-H.K., Soft docking: Matching of molecular surface cubes, J. Mol. Biol. 219, 97-102 (1991)

[3] Mitchell E.M., Artymiuk P.J., Rice D.W., Willett P., Use of techniques derived from graph theory to compare secondary structure motifs in proteins, J. Mol. Biol. 212, 151-166. (1989)

[4] Vriend G., Sander C., Detection of Common Three-Dimensional Substructures in Proteins, Proteins: Structure, Function and Genetics 11,52-58 (1991)

[5] Alexandrov N.N., Takàhashi K., Go N., Common Spatial Arrangements of Backbone Fragments in Homologous and Non-homologous Proteins, J. Mol. Biol. 225 5-9 (1992)

[6] Fischer D., Norel R., Wolfson H., Nussinov R., Surface Motifs by a Computer Vision Technique: Searches, Detection, and Implications for Protein-Ligand Interactions, PROTEINS: Str. Funct. and Genetics. 126, 278-292 (1993)

[7] May A.C.W., Johnson M.S., Protein structure comparisons using a combination of a genetic algorithm, dynamic programming and least-squares minimization, Protein Engineering. 7, (4) 475-485 (1994)

[8] Kuhl F.S, Crippen G.M., Friesen D.J., A Combinatorial Algorithm for Calculating Ligand Binding, J. Comput. Chem. 5 (1) 24-34 (1984)

[9] Boberg J., Salakoski T., Vihinen M., Selection of a Representative Set of Structures From Brookhaven Protein Data Bank; Proteins: Structure, Function and Genetics 14, 265-276 (1992)

[10] Subbaiah S., Laurents D.V., Levitt M., Structural similarity of DNA-Binding domains of bacteriophage repressors and the globin core, Current Biol. 3 (3), 141-147 (1992)

Chapter 5

FRACTAL MODELING AND IMAGE PROCESSING

SIMULATION OF GROWTH AND FORM OF SPONGES AND CORALS AND THE INFLUENCE OF THE PHYSICAL ENVIRONMENT

Jaap A. KAANDORP

Parallel Scientific Computing and Simulation Group, Faculty of Mathematics and Computer Science, University of Amsterdam, Kruislaan 403, 1098 SJ Amsterdam, The Netherlands

ABSTRACT

In this paper examples of 3D models with which the morphogenesis of a certain class of sponges and stony-corals can be simulated are presented. In these models the influence of the physical environment, consisting of the distribution of light intensities and nutrients in the environment, is included. The growth process of photosynthetic or filter-feeding marine organisms can be mimicked with these models. Possible applications are in bio-monitoring, where changes in the actual growth forms caused by changes in the physical environment, can be detected and interpreted using these simulation models.

RESUME

Dans cet article les exemples d'un modèle à 3 dimensions sont données. Avec ce modèle on présente la possibilité de simuler la morphogenèse d'une classe d'éponges et des corailles (Sceleractinia). L'influence de l'environnement physique, de la distribution des intensités de la lumière et des substances de nutrition, est inclue dans ce type de modèles. Avec ces modèles on peut simuler la croissance des organismes marins photosynthétiques ou hétérotropiques. Les applications potentielles pour ces modèles se situent au niveau de la détection et de l'interprétation des changements dans les formes de croissance causés par les modifications de l'environnement physique.

1 INTRODUCTION

The emergence of forms in the growth process of biological objects is one of the most fundamental problems in biology. Very few exact results are available about how forms emerge in a growth process. In almost all cases, it is not yet known how the genetic information is physically translated into the actual form. A growth process can be considered as an iterative process: the growth form of an object in a certain growth stage is also determined by the shape of the object in a previous growth stage. It is often not possible to predict the shape of the object at a certain time stage in an analytical way. The problem of morphogenesis is often irreducible and the only feasible way to obtain more insight into this process is by simulation.

Examples of relatively simple growth processes are found among bacteria (Fujikawa & Matsushita, 1991; Matsuyama & Matsushita, 1993), fungi (Obert, 1993), and sponges and corals (Kaandorp, 1994). Although the growth process of these organisms is relatively simple, a surprisingly large variety of complex forms develops in the morphogenesis.

In this paper a short outline is given of three types of morphological models with which the growth process of a certain class of sponges and stony-corals can be simulated (see for more details Kaandorp, 1994).

2 THE GROWTH PROCESS IN SPONGES AND STONY CORALS

In many marine sessile organisms (for example sponges and stony-corals) the growth form emerges in an iterative process in which layers of material are added on top of the preceding growth stages of the organisms. In a majority of these organisms the preceding growth stages remain unchanged. In a longitudinal section, parallel to the axis of growth, the layered structure is revealed, from which it is possible to distinguish the form of the organism in an earlier growth stage (see Fig.1).

A special type of growth process where a layered structure evolves is radiate accretive growth. In this type of growth process the longitudinal elements are set perpendicular to the preceding growth stages, while the surface of the organism is tessellated with tangential elements. The thickness of a newly secreted layer is greatest when the angle between the normal surface vector and the direction of the growth axis is 0°, while a zero or minimal thickness is attained at larger angles (in Fig.1 this angle is 90°). The consequence is that the highest growth velocities develop at the tips of the branches of the organism.

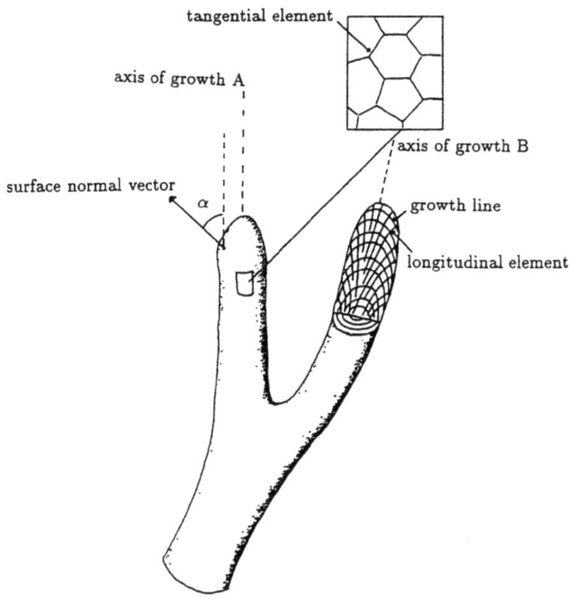

Figure 1. A diagram of an organism with radiate accretive growth. A part of one the branches is removed in order to reveal the layered structure of the organism. A part of the surface is enlarged to show the arrangement of the tangential elements (see inset).

Radiate accretive growth can be found among various marine sessile organisms, as for example stony-corals (Graus & Macintyre, 1982) and sponges (Kaandorp, 1991). The layers may consist of very different chemical materials, as for example calcium carbonate or silicium. The layers may be organized in polyps (stony-corals) or in spicules (sponges). Although the organisms are from very different groups and consist of very different materials, the skeleton reveals in a longitudinal section a generic structure as shown in Fig.1. The tangential and longitudinal elements in this figure may correspond with the edge of a corallite in a stony-coral or a spicule in a sponge

Sponges and stony-corals with radiate accretive growth use light, filter-feeding by trapping suspended material from the environment or a combination of both as an energy source. Where light is being used exclusively for the photosynthesis, only light and inorganic substances are required for the growth process. Where filter-feeding is the only energy input, the growth process is influenced by the local concentration of (organic) suspended material. In many of these organisms (e.g. many stony-corals) a combination of both energy sources is used. In the radiate accretive growth process a wide variety of growth forms may emerge. The growth process is strongly influenced by the physical environment and in some cases it is possible to arrange the growth forms along an environmental gradient.

In Fig.2 growth forms of the sponge *Haliclona oculata* with radiate accretive growth are shown. In the sequence A..C the thin-branching form A gradually transforms into a plate-like form C. The plate-like form is collected from a site exposed to water movement, the thin-branching from a sheltered site, while sample B originates from an intermediate site.

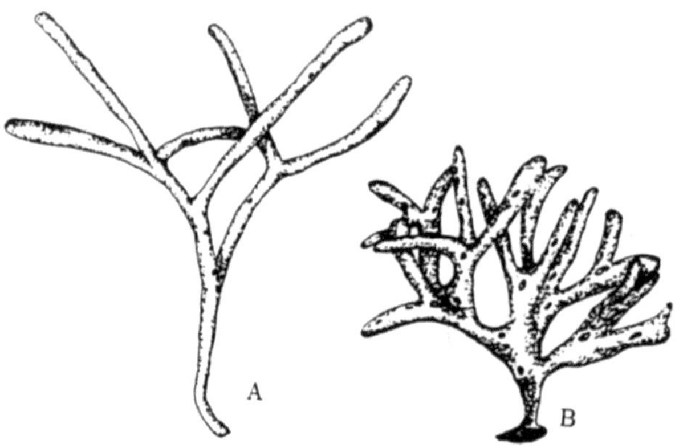

Figure 2. Range of growth forms of the sponge *Haliclona oculata* collected along an environmental gradient, in the range A..C the exposure to water movement increases.

3 The Model

The radiate growth process is simulated by an iterative geometric construction in which layers are constructed on top of the preceding iteration steps. In Fig.3 the construction of a new layer on top of the preceding layer is shown. The tangential elements are represented by edges of triangles organized in layers, while the longitudinal elements are represented as edges between the corresponding vertices of the successive layers. The iteration process itself starts with a triangulated sphere. In this triangulated sphere, and in the successive layers, the triangles are organized in pentagons and hexagons, which corresponds to the surface tessellation of an actual organism. The thickness of a new layer of material which is added in the iterative geometric construction, is represented in the model by the length l of a longitudinal element. The length l is determined by:

$$l = \begin{cases} s \,.\, growth\_function \,.\, h(\overline{r}, max\_curv) \\ \text{for } growth\_function \,.\, h(\overline{r}, max\_curv) > 0.02 \\ 0.0 \text{ for } growth\_function \,.\, h(\overline{r}, max\_curv) \leq 0.02\, s \end{cases} \quad (1)$$

The length l in Eq.1 is expressed in a constant s, representing the size of the basic building elements, the function *growth_function* and the function $h(r, max\_curv)$ are all within the range 0.0..1.0. In this equation the threshold 0.02 s is used for l. In the actual objects the longitudinal elements are constructed from discrete elements which cannot become arbitrarily short. Below this threshold no material is added, vertices in the geometric construction of Fig.3 with a longitudinal element with l above this threshold are indicated as active. The direction of the longitudinal element (V_1, V_2) is determined in the basic construction of 3, by the mean value of the normal vectors of the triangles that surround the vertex V_1 of the preceding layer. In the following subsections three types of growth functions ($f(\alpha, \beta)$, $I(\theta)$, and $k(c)$) and the "amount of contact with the environment function" $h(r, max\_curv)$ will be briefly discussed.

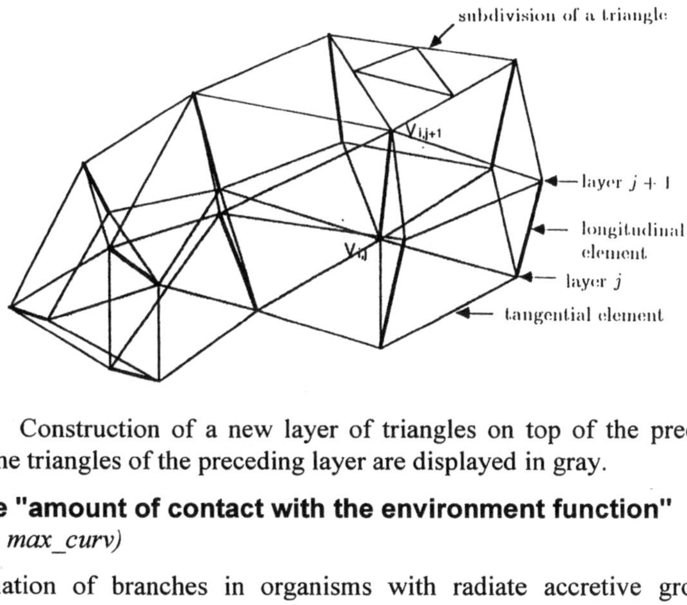

Figure 3. Construction of a new layer of triangles on top of the preceding layer. The triangles of the preceding layer are displayed in gray.

3.1 The "amount of contact with the environment function" *h(r, max_curv)*

The formation of branches in organisms with radiate accretive growth can be described as a self-organizing aspect in the growth process. The heterotrophic organisms with radiate accretive growth are filter feeders which capture suspended material from the environment. The growth velocities which develop in a newly added layer also depend on the capabilities of the transport system with which nutrients are transported through the tissue of the organism. The access to the suspended material in the environment can be described as a quotient of the surface through which nutrient can pass and the volume (with a certain distance from the surface) being supplied with nutrient. The amount of contact with the environment can be estimated by measuring the local radius of curvature in a vertex in several tangential directions on the surface of the object. In the actual organism the amount of contact with environment will decrease when the local radius of curvature increases. The amount of contact with the environment is expressed in the function *h(r, max_curv)*. In this function the argument *r* is the mean value of the local radii of curvature. When a certain critical value, the maximum radius of curvature *max_curv,* is exceeded that is related with the capabilities of the transport system, no nutrient at all will be transported in the actual organism and in the model *h(r, max_curv)* will become zero. For lower values of *r* this function returns values above zero. Under circumstances where the transport system is supported by external water movement as well, the value of the model parameter *max_curv* will be relatively higher.

3.2 The empirically derived growth function $f(\alpha, \beta)$

The growth function $f(\alpha, \beta)$ is derived by measuring the thickness of successive growth layers and the angle between the normal surface and the direction of the growth axis using an actual longitudinal section as shown in Fig.1. In the isotropic case the empirically derived growth function as measured in a longitudinal section will always show about the same, as long as the section is made parallel along the axis of growth. In many cases the growth process is anisotropic and the empirical growth function shows a stronger widening effect in the plane perpendicular to the governing flow direction; this widening effect decreases when the section is made in a plane with a smaller angle with the flow direction. The emergence of a flattened growth form is simulated with an anisotropic growth function in which the growth velocity also depends upon the angle β. When the flow direction is assumed to be parallel with the yz-plane, β is defined as the angle between the yx-plane and the projection of the normal surface vector on the xz-plane. This function has the largest plateau of maximum values when β = 0; the widening decreases towards a minimum when β = π/2. Two examples of empirically derived growth functions are depicted in Fig.4.

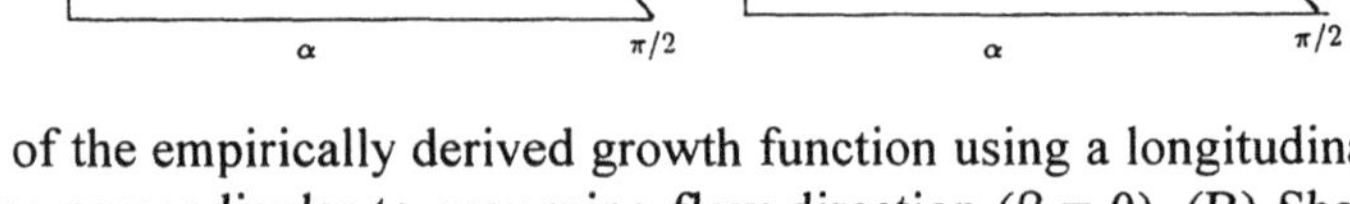

Fig. 4. Shape of the empirically derived growth function using a longitudinal section made in a plane perpendicular to governing flow direction (β = 0). (B) Shape of the empirically derived growth function using a longitudinal section madein a plane with an angle β = π/2. The angle α in (A) and (B) is the angle between the surface normal vector and the axis of growth, *l* is the normalized distance between two successive growth layers.

When $f(\alpha, \beta)$ is used as a growth function in Eq.1, flattened objects are generated, as shown in Fig.5.

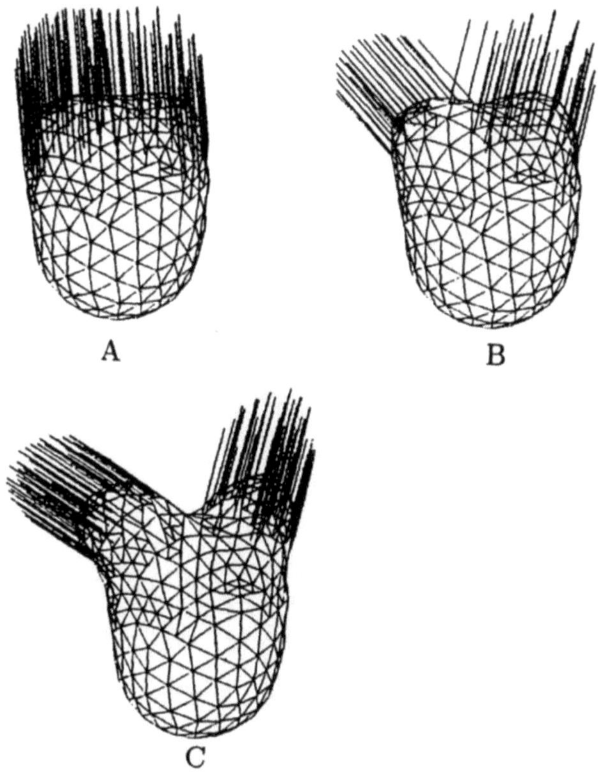

Figure 5. Objects subsequently emerging in the iteration process, in each active vertex the direction of the growth axis is displayed with an edge. After a fewiteration steps the original patch of active vertices is separated into two newones; both patches are associated with different growth axes.

In the first object all active vertices are associated with the same growth axis. In the active vertices the length l of the new longitudinal element has not dropped below the threshold 0.02s in Eq.1. The active vertices represent the sites in the actual object where growth still takes place. After some iteration steps a local minimum develops and the patch of active vertices is separated into two new ones, because locally the maximum allowed radius of curvature *max_curv* is exceeded and r decreases. In the two new patches the growth process continues independently and two new growth axes develop. A new growth axis in a patch is defined by the direction of the longitudinal element with a length l that is a local maximum. In Fig.6 a few more evolved examples are shown generated with the growth function $f(\alpha, \beta)$.

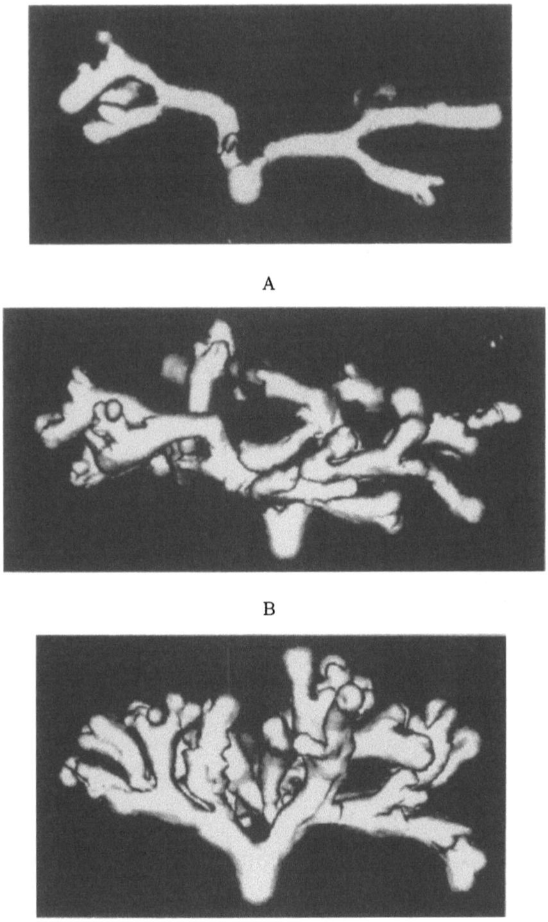

A

B

C

Figure 6. Three objects generated with the iterative construction shown in Fig.3 and the growth function *f*(α, β). In the series A..C the parameter *max_curv* increases.

3.3 The influence of the local concentration gradient of suspended material (function *k(c)*)

The influence of the local concentration gradient of suspended material on the growth process can be represented by using an estimation of this gradient, function *k(c)*, as a growth function in Eq.1. The local concentration gradients are determined by using a lattice version of the simulated object. The object is first mapped onto a lattice, for this purpose in the simulations of the heterotrophic organisms a lattice of 100^3 sites is used, where each lattice site contains a concentration value in the range 0.0 .. 1.0. After this a solution of the Laplace equation is approximated, and the concentration for each lattice site and the gradients are estimated. In the approximation it is assumed that the concentration on the object itself is zero. The lattice sites situated on the surface and in the inside of the object, as well as the lattice sites at the bottom plane of the lattice are set to the value zero. In the approximation a linear source is assumed by setting the lattice sites of the top plane to the value 1.0. In Fig.7A a more evolved example is shown generated with the growth function *k(c)*.

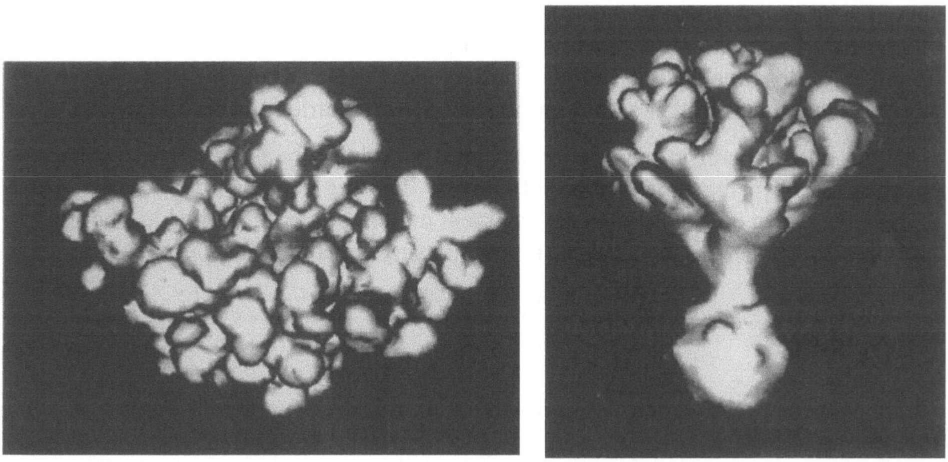

A B

Figure 7. Objects generated with the iterative construction shown in Fig.3. In (A) the growth function *k(c)* was used, while (B) was created using the function $I(\Theta)$.

3.4 The influence of the local light intensity (function $I(\Theta)$)

The influence of the local light intensity on the growth process of the phototrophic organisms is modeled using a growth function $I(\Theta)$, which represents the local light intensity at the surface of the object. In the determination of the local light intensity on the surface of the object a lattice version of 400^3 sites of the geometric model is used. The local light intensity is calculated in each triangle using the cosine of the angle of incidence Θ between the normal vector of the triangle and the light source. In the simulations a parallel light source is assumed and the light direction corresponds with the vertical. The lattice version is used to estimate the part of each triangle that is shaded by parts of the object situated above the considered triangle. After the local light intensities are corrected for this shading effect, $I(\Theta)$. is used to determine l. In Fig.7B a more evolved example is shown generated with the growth function $I(\Theta)$.

Conclusion

With the growth function $f(\alpha, \beta)$ it is possible to generate a range of forms which gradually change from a thin-branching form (Fig.6A) into a plate-like form (Fig.6C). This range of forms corresponds to the range of actual growth forms shown in Fig.2. It is possible to make the model physically more correct by introducing the influence of the distribution of nutrients in the environment on the growth process by using the growth function $k(c)$. With this approximation the concentration distribution of suspended material around a filter-feeding organism under sheltered conditions is imitated. To describe the nutrient distribution under exposed conditions it is necessary to extend the model further and to introduce a flow of nutrients in the model.

A model of an organism using phothosynthesis exclusively as an energy source can be obtained by using the function $I(\theta)$ in the iterative construction. Models which include the effect of the physical environment may in future be applied to trace and to interpret changes in the physical environment using actual growth forms.

Keywords

Morphogenesis, simulation of growth and form, stony-corals, sponges, bio-monitoring.

References

[1] Fujikawa H., Matsushita M., Bacterial Fractal Growth in the Concentration Field of Nutrient, in J. Phys. Soc. Japan, 60(1) 88-94 (1991).

[2] Graus R. R., Macintyre I. G., Variation in Growth Forms of Reef Coral Montastrea nnularis (Ellis and Solander) : a Quantitative Evaluation of Growth Response to Light Distribution using Computer Simulation, Smithson. Contr. Mar. Sci., 12:441-464 (1982).

[3] Kaandorp J. A., Modeling Growth Forms of the Sponge Haliclona oculata (Porifera ; Demospongiae) using Fractal Techniques, Mar. Biol., 110:203-215 (1991)b.

[4] Kaandorp J. A., A Formal Description of Radiate Accretive Growth, J. Theor. Biol., 166:149-161 (1994).

[5] Kaandorp J. A., Fractal Modeling : Growth and Form in Biology, Springer- Verlag, Berlin, New York (1994).

[6] Matsuyama T., Matsushita M., Fractal Morphogenesis by a Bacterial Cell Population, Critical Reviews in Microbiology, 19(2):117-135 (1993).

[7] M. Obert., Microbial growth patterns: fractal and kinetic characteristics of patterns generated by a computer model to simulate fungal growth. Fractals, 1:354-374 (1993).

FRACTAL SHAPE DESCRIPTION USING COMPACT DATA SETS

Huw JONES

Reader of Computer Graphics, Centre for Electronic Arts, Middlesex University, Barnet EN4 8HT, UK, E-mail: huwl@mdx.ac.uk

ABSTRACT

The concept of a 'fractal object' is discussed and two methods for generating fractals are described, illustrating how complex objects can be generated from small data sets. A method for fractal compression of image data is outlined, and possible extensions of this to object modeling are indicated.

RESUME

Le concept d' "objet fractal" est discuté et deux méthodes pour générer des fractales sont décrites, illustrant la génération d'objets complexes à partir d'ensembles restreints de données. Une méthode de compression fractale de données d'images est esquissée et on aborde le problème de son extension possible à la modélisation d'objets.

1 INTRODUCTION: FRACTAL CONCEPTS

The term 'fractal' was coined by Benoit Mandelbrot during the 1970s. The precise definition of a fractal object involves some concepts of point set topology which are fairly obtuse to the lay person:

'A fractal object is by definition a set for which the Hausdorff Besicovitch dimension strictly exceeds the topological dimension.' [Mandelbrot 1983, plS].

Mandelbrot himself seems uneasy about using such a definition, 'I continue to believe that one would do better without a definition' [Ibid, p361]. It has proved fruitful to approach an understanding of fractal objects through considering their defining secondary properties, which are

- Self similarity: subsets of the object are, in some sense, copies of the whole;
- Infinitesimal subdivisibility: there is no apparent change in the amount of detail observed at different levels of magnification.

Fractal objects have been investigated in mathematics since at least the later stages of the 19th century, long before the word was coined. Point set topologists, such as Peano, Cantor, von Koch and Sierpinski developed creations with apparently paradoxical properties that created a split between them and traditionalists.

The Koch curve and Sierpinski triangle (Fig 1) are examples of such constructions. The Koch curve's length is multiplied by 4/3 at each stage of construction, so has infinite limiting length [Jones 1991, pl2], but the area below it is finite. As it contains infinitesimally small copies of itself, any ' small' section of it will also be infinite in length. The Sierpinski triangle (or gasket) can be constructed by recursive drawing of small triangles within larger triangles, or as a single curve [Sierpinski 1916] which, like the Koch curve, has no properly defined tangent. It is easy to show that the Sierpinski triangle has zero area, although it is not empty.

Such objects do not seem to possess a natural dimension. They are somehow more than curves, yet they do not fill a two dimensional space. If we take a more 'normally' dimensioned object that can be subdivided into N subcopies of itself at scale f (N = 4 when f = 0.5 for a square in 2D), its dimension satisfies

$$N = (1/f)^D.$$

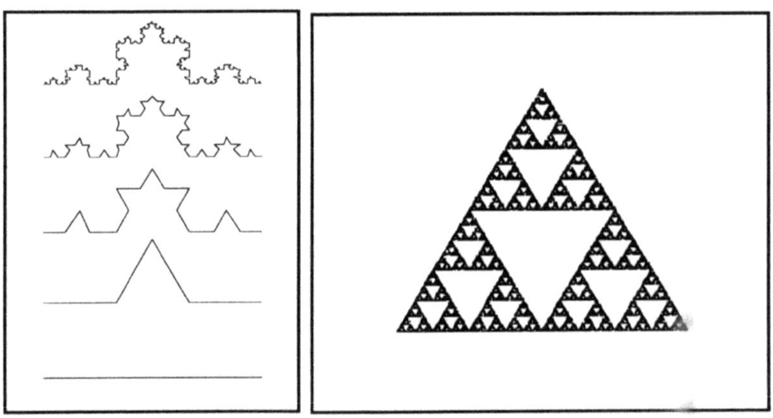

Figure 1. Generation of a Koch curve and a Sierpinski triangle.

This law can be applied to fractal objects which are exactly self similar to obtain a fractal dimension (or Hausdorff dimension) [Voss 1988, pp 28-9] as

$$D = \log(N)/\log(1/f).$$

Substituting N = 4 and f = 1/3 for the Koch curve and N = 3 and f = 1/2 for the Sierpinski triangle gives fractal dimensions of 1.2619 and 1.5850 for these forms respectively.

Examples given above are '*exactly self similar*' fractals, in that they contain precise subcopies of themselves. Other fractal objects are '*statistically self-similar*', with subparts having similar, although not identical, forms to the whole.

Irregular coastlines are observed to have such properties. At scales above the sand grain level, it is difficult to distinguish the scale of a coastline from its outline alone. Fractal dimensions are found using a neighbourhood counting technique. The object is 'covered' by neighbourhoods of particular sizes - for our purposes, neighbourhoods can be taken as small circles or spheres depending on the environment's dimension.

If N_1 neighbourhoods of a particular size is the minimum number needed to cover the object, and N_2 are needed when the neighbourhood size is scaled by a factor f, then the dimension is the limiting value of

$$\log(N_2/N_1)/\log(1/f)$$

as the neighbourhood size approaches zero. Figure 2 illustrates how the fractal dimension of a curve can be estimated. To cover a non-fractal curve, about twice as many half sized neighbourhoods would be needed. When the curve has fractal properties, relatively more of the small neighbourhoods are needed.

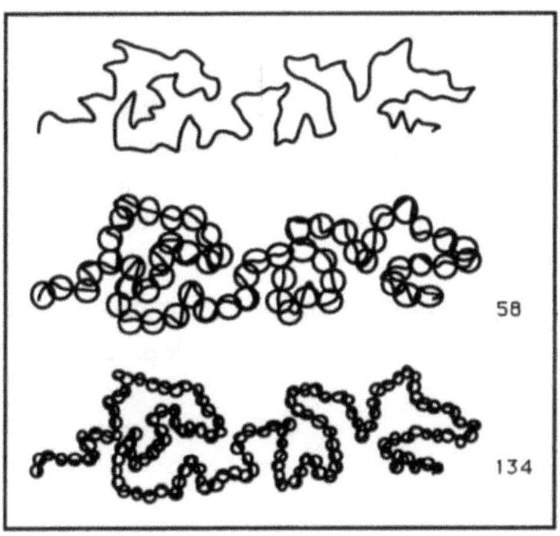

Figure 2. Curve has fractal dimension D=log(134/58)/log(2)=1.21(to 2dp)

2 L-SYSTEMS FOR COMPUTER GENERATION OF FRACTAL OBJECTS

There are many methods for generation of fractals. We mention two which illustrate the recursive methods used for fractal generation and show how complex structures can be encoded by relatively little data. L-systems (named after Aristid Lindenmayer, who developed them to model botanical forms) [Prusinkiewicz 1990] consist of an alphabet (a set of characters), an axiom (a starting string of characters) and productions (a set of rules for converting characters to strings). Productions are applied to all characters of the existing string simultaneously and the process is repeated until the required level of recursion is achieved. Typically, characters are interpreted as drawing commands as in LOGO's turtle graphics to produce an image. Using F to represent a forward line, + to denote a right turn of angle δ and - to denote the equivalent left turn, the Koch curve (fig 1) can be encoded with $\delta = 60°$ using

Axiom	F
Production	F → F-F++F-F

Characters for which no rule is given are unchanged and the length of an 'F' is reduced by a factor of one third each time a production is applied. By including bracket characters [and] to indicate starts and ends of branches (implemented by use of a stack), with branching angles also defined, more complex branching structures can be developed. Figure 3 shows the first four generation stages of the simple L-system with $\delta = 45°$.

Axiom	F
Production	F → F[+F-F]FF

More complex objects can be produced by allocating parameters to the drawing commands, introducing stochastic elements to productions and the drawing phase, making productions 'context sensitive', in which the string neighbours of a character affect the application of a production, and interpreting strings as objects in higher dimensional spaces.

Many simulations of natural phenomena are given in Prusinkiewicz [1990]. Structures developed by this method are closely related to the developments shown by Kaandorp [1993].

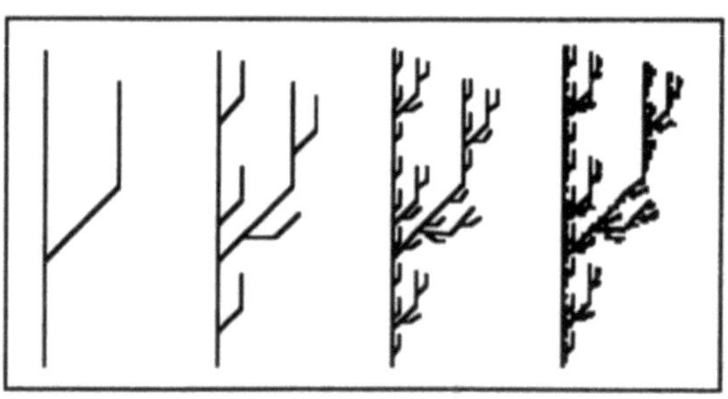

Figure 3. 4 stages in generation of an L-system.

3 Iterated Function Systems for Computer Generation of Fractal Objects

Iterated Function System (IFSs) have been popularised by Michael Barnsley since the mid 1980s, but the work of Williams [1971] and Hutchinson [1981] are cited as pioneering publications. An IFS consists of a set of contractive affine transformations $\{T_i\}$ [Barnsley 1993]. Such transformations preserve parallel lines and can be represented as a combination of translations, contractive scalings and rotations. If X is a column vector representing a Cartesian point, then

$$T_i(X) = A_iX + B_i,$$

where A_i is a square matrix and B_i is a column vector of constants, is an affine transformation. The transformation is contractive if only and if the distance between any two distinct points in the transformed figure is less than the distance between the equivalent points in the original figure. Affine transformations are defined in 2D by 6 numbers, in 3D by 12 numbers (those required to fill A_i and B_i). An IFS will have a single 'attractor', an image or object which remains unchanged when all transformations comprising the IFS are applied to it, so the IFS can be taken to represent this object. Each transformation maps the whole object to a part of itself that is an affine transformed copy of the whole.

A three transform IFS with

$$A_1 = A_2 = A_3 = \begin{bmatrix} 0.5 & 0 \\ 0 & 0.5 \end{bmatrix}, \quad B_1 = \begin{bmatrix} 0 \\ 0 \end{bmatrix}, \quad B_2 = \begin{bmatrix} 0.5 \\ 0 \end{bmatrix} \quad \text{and} \quad B_3 = \begin{bmatrix} 0.25 \\ \sqrt{3}/4 \end{bmatrix}$$

represents a Sierpinski triangle (Fig 1, right). The three transformations give ways of producing the three half scale copies from the original. As all the scaling matrices A_i are identical and uniformly scaled, there are even more compact ways of describing this object.

Four rectangles in the left of Fig 4 represent four contractive affine transformations of the surrounding square.

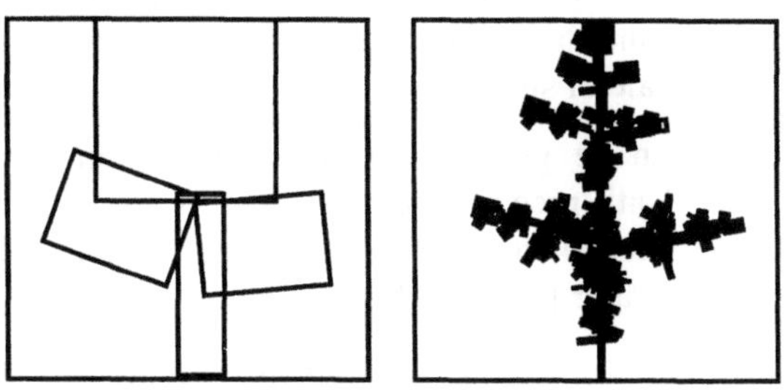

Figure 4. Four transformation of an IFS and their attractor.

Figure 4 (right) shows their attractor, generated by recursively applying the four transformations to an original black square (any shape could be used as the original form). The attractor stabilises after about 5 iterations, showing how the relatively simple data of the left image encodes the complex image of Fig. 4 (right).

Barnsley's "chaos game"

This is an alternative generation method. Each of the transformations is allocated a non zero probability such that all probabilities sum to unity. Given a starting point, select a transformation at random using this probability distribution, apply this transformation to the starting point and plot the resulting point. Repeat this process, selecting a new transformation and re-using the plotted point of the previous phase as the new starting point.

This method is impractical using pencil and paper, but using a computer, the image will emerge from an originally blank screen in a relatively short time. Given an arbitrary starting point, the first 25 or so points should not be plotted, but the remaining points should have converged into the required shape. The allocated probabilites are arbitrary, as the same shape will eventually be generated, but if probabilities are proportional to the area contractions of their transformations, image generation will be of even density throughout the generation. The method can be extended to objects lying within 3 dimensions.

IFS structure and objects

Object depiction within 3D can be aided using the z-buffer and depth cueing algorithms from computer graphics. Jones and Campa [1993] show how classical abstract forms such as the Sierpinski tehahedron and Menger sponge can be generated by this method, develop some new abstract forms and model natural phenomena such as clouds and trees. By limiting possible inter-transform steps, their technique is similar to the recurrent iterated function systems described by Hart [1994a]. They also show that IFSs can depict non-fractal forms, such as a tetrahedron.

Hart and deFanti [1991] imposed pseudo surface normals on IFS generated fractal structures to aid visualisation. Abstract fractal objects do not have well defined normal surfaces, so their use in this context is debatable, although the method is perfectly valid when showing simulations of natural objects whose fractal structure only holds down to a limited scale of subdivision.

A wide variety of objects can be generated using IFSs which encode them by relatively compact data sets. Until recently, some degree of '*artistry*' was needed to develop the IFS structure needed to depict particular forms, often involving interactive adaptation by an operator [Horn 1991].

The '*collage theorem*' describes how '*tiles*' (possibly overlapping) which cover the required object with affine copies of itself give transformations that define an IFS [Barnsley 1993]. The '*inverse problem*' of IFSs was stated by Barnsley et al [1986], 'given an object, find an iterated function system that represents that object within a given degree of accuracy.' This is the central problem for fractal image compression.

4 FRACTAL IMAGE COMPRESSION

A team working with Michael Barnsley has generated much interest in this topic. There have been ambitious claims for image compression ratios - 'the ratio of the memory required to hold an image as a set of pixels to the memory required to store a representation of the image in compressed form' [Fisher 1992] - from Barnsley's with his coworkers. Compression rates 'exceeding 10 000 to 1' have been claimed for images created by Barnsley's students, although the method is 'computationally-intensive'.

Barnsley's forms

A later release from Barnsley's company Iterated Systems, Inc. [1992] gives a typical storage requirement of 10Kb for a standard colour television image compared to the uncompressed size of 768Kb; this is still an impressive ratio.

Early claims led to considerable interest; the bandwidth requirements for television images will be greatly reduced if compressed images can be transmitted. However, 'in spite of the popularity of the notion, scientific publications on the topic have been sparse; most articles have not contained any description of results or algorithms.

Even Barnsley's book .. was spartan when it came to the specifics of image compression.... a student ... published related papers with Barnsley without revealing their core algorithms' [Fisher 1992, p903]. Arnaud Jacquin's PhD thesis [1989] gave details of the method, further explained in journal papers such as Jacquin [1992], followed by Barnsley and Hurd [1992].

Hart [1994b, p9.2] states that methods for automatic fractal modelling of objects 'have not yet become practical' and, of earlier applications of the method, 'their less than spectacular results were offset by its exciting potential.'

Most questions concern the time taken in the compression stage. Given enough lead time, images can be compressed impressively, but real time compression is the ultimate aim for television. Barnsley's company, Iterated Functions, Inc. has developed image encoders and decoders as commercial products.

Fractal Encoding

Saupe and Hamzaoui [1994] review related literature and give ftp sites for three public domain programs as well as details of the commercial software. Fisher [1994] gives as complete as possible a view of the current state of fractal compression schemes, and gives C code 'capable of encoding images in a few seconds, decoding at arbitrary resolution and achieving high compression ratios,' so the situation is improving.

Pixellated forms

Yuval Fisher [1992 pp 904-919, 1994 pp 55-78] gives two very clear descriptions of fractal image compression methods, the latter discussing compression times and ratios in detail. A brief description of the methods used is given below.

The image to be compressed is defined in pixellated form, as it would appear on a television screen. Pixels (picture elements) are grid addressed locations that hold the intensity of tone displayed at the grid location. A colour image contains three monochrome images, displayed as (red, green, blue) triplets in the additive colour space of a television image, where light is added to the environment, or (cyan, magenta, yellow) in the subtractive colour space of the printed page, where light is filtered out of the environment.

For high quality colour printing, a black component would also be added, but these are complexities that do not affect the underlying principle of the method. Intensities in pixel based images can be considered as discretised or sampled versions of an underlying image intensity i(x,y) defined on the continuous Cartesian plane, having interesting consequences for the reconstruction algorithm.

Fractal self similarity

Fractals possess '*self similarity*'. Fractal compression identifies range portions of the image that are similar to contractive affine transformed versions of larger domain portions. If ranges can be represented as affine contractive transformations of a domain, using only six numbers for locational transformation and two numbers for colour space transformations in the monochrome case, considerable image compression can take place.

Ranges and domains can be found using a 'quadtree' subdivision of the image space, where the image is repeatedly divided into four quarters, so it is easier to work with images based on a 2^k square, although algorithms can be adapted to cope with any shape of image. At each stage, unallocated range squares can be compared with domain squares from the previous stages to search for an adequate 'fit', the process repeated until the whole image is covered with matched up range squares or a given level of subdivision is reached, when a 'best possible' match is accepted.

An affine relationship of the form

$$i_r = s i_d + o,$$

between the levels of intensity i_r displayed in individual pixels of a small range square with mean intensities i_d of equivalently located pixel clusters in a larger domain square is sought. The constants s (scaling) and o (offset) may be estimated by least squares.

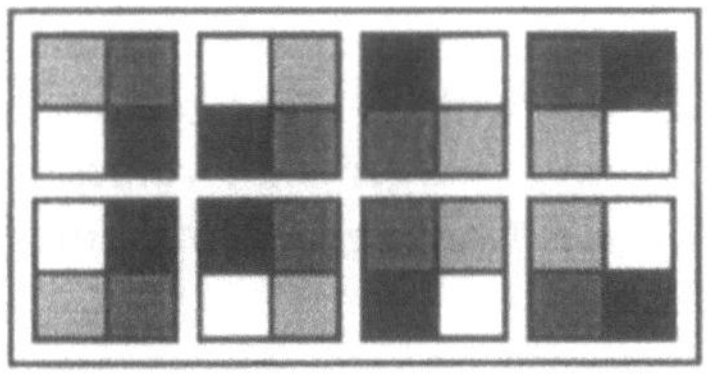

Fig.5 Eight affine transformations of a square

The transformation is contractive when s is numerically less than 1. There are eight possible spatial orientations that map the domain to the smaller range, taking into account four rotations and two reflections (Fig 5).

A direct method would compare all eight orientations of a range with a domain, but obvious non-fits can be eliminated by testing only squares which have a matching arrangement of intensities from dark to light squares. There are only three such arrangements (Fig 6), providing considerable time saving.

Constants s and o used in this transformation are held to limited precision. Fisher [1994] uses five bits for s and seven bits for o to give acceptable results in terms of image precision whilst giving good compression ratios), so the limited precision values should be used in finding a measure of difference between range and transformed domain.

Maximum absolute difference or mean square measures can be used to measure these differences. If a difference lies within a given precision, the range square can be represented by a three bit code identifying the orientation match between range and domain, twelve bits for s and o and six 'short' integers identifying range and domain locations.

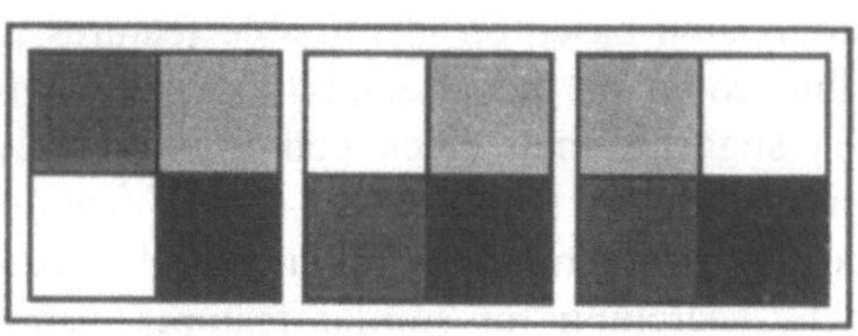

Fig.6 Three possible gradations of intensity for four sections of a square

When the surface of the image is covered by non-overlapping domains which are mapped to larger ranges, the compression is complete. IFSs involve sets of *contractive* affine transformations. Geometric transformations from domain to range using these methods are necessarily contractive. In searching for an image intensity match, the mapping is a contractive transformation if the factor s is numerically less than one. Fisher [1994, pp62-63] indicates that maximum permissible numerical values of s can be increased to as large as four with the repeated application of all transformations remaining eventually contractive. Larger limit values for s can give greater image precision at the potential cost of slower regeneration times. Algorithms to counter this effect can be implemented.

Image Regeneration

Regeneration involves recursively applying the complete set of affine transformations (the IFS) to an arbitrary original image such as a plain black frame until differences between successive generations are acceptably small, as in the IFS of Fig. 4, or by the 'chaos game algorithm', which fills the frame with a random cloud of points that eventually complete the image. Hart [1994b, pp 9.7 and 9.10] shows how four iterations of the former method give recognisable image information, with 512 iterations of the chaos game giving similar precision. Fisher [1992, p913] shows how encoding can be represented in matrix form, reducing the limiting form of decoding to a single multiplication by a sparse matrix inverse. If fast algorithms are available for the matrix manipulation, this gives a good decoding technique. Fisher [1992, p916-7] shows a convincing quadtree compressed image of a dog at about 30:1 compression ratio with rms error of 8.5 on an 256 level intensity. An image of a cityscape with fine detail is compressed at a ratio of 7.6: 1 using a rectangle subdivision method which searches for suitable rectangle domains to range matches, yielding rms error of 7.1. Postprocessing which averages pixel intensities across range boundaries improves the visual continuity of the decompressed image. Compression ratios and noise have to be traded. For a particular algorithm, better image quality can only be achieved at lower compression ratios.

Image subdivision

Algorithms are available to compress according to a given image integrity or compression ratio. Triangular subdivision based on identifiable features is proposed as a potential method of avoiding some of the artefacts observed in rectangular subdivision. Hart [1994b, p9-11] suggests that 'block coding' could give way to pattern recognition methods which positively seek out self-similar features that 'not only compress data, but reveal its structural self-similarity'. IFS based fractal compression encodes an image by searching for similar features, using the self-similarity property of fractals.

Image regeneration at any scale

The '*infinitesimal subdivisibility*' property of fractals means that an image can be regenerated from such a compression *at any scale,* even at higher precision than the original. The limiting form of the regeneration is created as a continuous function on the image domain.

Additional 'detail' thus produced has no validity, so such images must be interpreted with care. If a pixellated image is enlarged, pixels become visible as square regions of constant colour. If such an image is IFS compressed, then regenerated at the larger scale, such artefacts are smoothed out, giving a more pleasing visual appearance. The extra spurious detail so created may partly explain some of the large image compression ratios claimed for early systems.

5 POTENTIAL FOR DEPICTION OF OBJECTS IN OTHER DIMENSIONS

The methods described above treat image intensity as a function i(x, y) of two Cartesian spatial coordinates. There is no difficulty in reducing this to a univariate problem, so continuous or discrete time series data could be condensed by this method. Television image information is transmitted in this form, so fractal compression could be applied directly to the signal as a time series. This may be less productive than the 2D form, as the two dimensional coherence of the image will not be used as a contributory factor. In the context of certain forms of music having fractal properties [Hsu 1993], it could be interesting to compress existing musical forms, regenerating them at arbitrary scales.

Voxel based model of 3D objects

There may be potential for using a three variable version of the method to depict voxel based models of 3D objects. A voxel is a 3D analogue of a 2D pixel. The space in which an object of interest lies is divided into a 3D grid of small cubes. The spatial occupancy of the object is defined to the limited precision of the grid size by identifying all cubes lying inside it [Kaufman 1991].

This method was first considered over 20 years ago, but was sidelined as being too memory intensive for existing equipment, although only a single bit was used to describe each voxel in a 3D Boolean array. Using an octree subdivision of space (a 3D version of the 2D quadtree image division described above) and by adding surface information to boundary voxels [Navazo et al 1986], some of the memory and precision deficiencies were ameliorated.

The enhanced computer memory now available at relatively little cost has signalled a resurgence in use of voxel based methods. Intensities held at each voxel represent, for example, density of the material at that location.

This has been used to model complex objects such as the human body using information from CAT scans or similar methods [Trousset and Schmitt, 1987]. Using high powered workstations, doctors can view the data from unusual angles, slice through the body to obtain novel views or strip away skin, muscle or organ tissue down to skeletal level by invoking relevant density thresholds. Models of this type are used for surgical training using virtual reality and have potential for use in remote surgery [Satava 1992]. An expert remotely situated surgeon may be called on to help with diagnosis, so the large data sets involved may need to be transmitted electronically.

Voxel images and decomposition integrity

Compression of the above medical data is potentially beneficial, and it is relatively easy to imagine how fractal image compression applied to pixel intensity images could be extended to accommodate voxel intensity descriptions of objects. The decompressed objects used in such circumstances need to be reconstituted with great integrity to ensure that important features are not missed and that spurious artefacts created by the compression-decoding sequence are not interpreted as actual elements of the original object. If IFS methods are to be used, the guaranteed precision version of a compression algorithm should be invoked, giving potentially lower compression ratio, and object reconstruction should not be done at any precision greater than the original object's voxel precision, to avoid the issue of spurious detail discussed above in the image case.

Archiving and data compression

There may be scope for compression in archiving information on object models in the medical case or in industrial situations where large catalogues of object definition data are held. Object modelling using Boundary Representation or Constructive Solid Geometry methods are well advanced for industrial applications and industrially created objects tend to have a limited range of discrete densities and require high spatial data integrity, so it is doubtful whether fractal compression methods would be of benefit.

The sequence of still frames that constitute a video image could be considered as a 3D block with the frame sequence adding the third dimension. Beaumont [1991] created domain blocks of 12 by 12 pixels by 12 frames with range blocks of 4 by 4 pixels by 4 frames. The method gave 'impressive' compression, but poor picture quality involving visual flicker probably due to the persistence of effects across the time variable. Enhancements to the method, involving selection of highly detailed sections for more intense treatment, produced improved effects and Barnsley's company, Iterated Functions Inc. has shown five minutes of the feature film *Casablanca* (Turner Entertainment Co.) generated from a compressed version held on a 1.44 Mb floppy disc.

CONCLUSION

Extravagant claims were made for fractal image compression when it was first proposed. More realistic compression ratios are achieved as the method matures, and it is seen to have considerable potential. Image compression can still not be performed in real time, but improved algorithms are likely to achieve that aim in a few years time, so real time transmission of fractally compressed television images will become possibile. Pre-prepared compression of still images, for example for transmission over the 'world wide web' could currently improve transmission time and limit required bandwidth, when slight losses in image integrity are acceptable. Equivalent methods in a 3D domain may prove to be beneficial in compressing object data definitions, but no great body of work has appeared in this area.

KEY WORDS

Fractal, image compression, iterated function system (IFS), L-system.

REFERENCES

[1] Barnsley M.F., Ervin V., Hardin D., Lancaster J., Solution of an inverse problem for fractals and other sets , Proceedings of the Royal Society A, 399, 243-275 (1986)

[2] Barnsley M.F., Hurd L.P., Fractal Image Compression, A.K.Peters, Wellesley, USA (1992)

[3] Barnsley M.F., Fractals Everywhere (2 ed), Academic Press, Boston, USA (1993).

[4] Beaumont J.M., Image data compression using fractal techniques, BT Technological Journal, 9(4), 93-109 (1991).

[5] Fisher, Y., A discussion of fractal image compression, in Peitgen H.-O., Jurgens H., Saupe D, Chaos and Fractals: New Frontiers of Science, Springer-Verlag, New York, USA, 903-919 (1992).

[6] Fisher Y., Fractal Encoding - Theory and Applications to Digital Images, Springer-Verlag, New York, USA (1994)

[7] Hart J.C., deFanti T.A., Efficient antialiased rendering of 3-D linear fractals , Computer Graphics, 25(3) (1991)

[8] Hart J.C., Iterated Function Systems and Recurrent Iterated Function Systems, New Directions for fractal modeling in computer graphics, Course Notes 13, ACM SIGGRAPH'94, Orlando, Florida, 1.0-1.22 (1994a)

[9] Hart J.C., Fractal image compression and the inverse problem of recurrent iterated function systems , New Directions for fractal modeling in computer graphics, Course Notes 13, ACM SIGGRAPH'94, Orlando, Florida, 9.1- 9.13 (1994b)

[10] Horn A., IFSs and the interactive design of tiling structures , in Fractals and Chaos, Crilly A.J., Earnshaw R.A., Jones H. (eds), Springer-Verlag, New York, USA, 119-144, (1991)

[11] Hsu K.J., Fractal geometry of music: from bird songs to Bach, in Crilly A.J., Earnshaw R.A., Jones H. (eds), Applications of Fractals and Chaos, Springer-Verlag, Berlin, Germany, 21-39 (1993)

[12] Hutchinson J., Fractals and self similarity, Indiana University Journal of Mathematics, 30, 713-747 (1981).

[13] Iterated Systems, Inc. Fractal Image Compression Technology, publicity release, Reading, UK (1992).

[14] Jacquin A.E., A Fractal theory of iterated Markov operators with applications to digital image coding, PhD Thesis, Georgia Institute of Technology, USA (1989)

[15] Jacquin A.E., Image coding based on a fractal theory of iterated contractive image transformations, IEEE Trans Image Processing, 1, 18-30 (1992).

[16] Jones H., Fractals before Mandelbrot - a Selective History, in Crilly A.J., Earnshaw R.A., Jones H. (eds), Fractals and Chaos, Springer-Verlag, New York, USA, 7-33 (1991).

[17] Jones, H, Campa, A. Abstract and Natural Forms from Iterated Function Systems, in Magnenat Thalmann N, Thalmann D (eds), Communicating with Virtual Worlds, Springer-Verlag, Tokyo, Japan, 332-344 (1993)

[18] Kaandorp J.A., 2D and 3D Modelling of Marine Sessile Organisms , in Crilly A.J., Earnshaw R.A., Jones H. (eds), Applications of Fractals and Chaos, Springer-Verlag, Berlin, Germany, 41-61 (1993)

[19] Kaufman A., Introduction to volume synthesis , in Patrikalakis N.M. (ed), Scientific Visualization of Physical Phenomena, Springer-Verlag, Tokyo, Japan, 27-35 (1991)

[20] Mandelbrot B.B., The Fractal Geometry of Nature, W.H.Freeman, New York, USA. (1983)

[21] Navazo I., Ayala D., Brunet P., A geometric modeller based on the exact octree representation of polyhedra , Computer Graphics Forum, 5(2) 89-104 (1986)

[22] Prusinkiewicz P., Lindenmayer A., The Algorithmic Beauty of Plants, Springer-Verlag, New York, USA (1990)

[23] Satava R.M. Robotics, telepresence and virtual reality: a critical analysis of the future of surgery, Minimally Invasive Therapy, 1, 357-363 (1992)

[24] Saupe D., Hamzaoui R, A guided tour of the fractal image compression literature , New Directions for fractal modeling in computer graphics, Course Notes 13, ACM SIGGRAPH'94, Orlando, Florida, pp 5.1-5.21 (1994)

[25] Sierpinski W., Sur Une Courbe dont tout point est un point de ramification , in W.Sierpinski: Oeuvres Choisis, Tome 11, PWN - Polish Scientific Publishers, Warsaw, Poland, (1916, republished 1975)

[26] Trousset Y., Schmitt F., Active ray tracing for 3D medical imaging , in G Maréchal (ed), Eurographics'87, North-Holland, Amsterdam, Netherlands, 139-150 (1987)

[27] Voss R.F., Fractals in Nature: from Characterisation to Simulation, in Peitgen H.-O., Saupe D. (eds), The Science of Fractal Images, Springer-Verlag, New York, USA, 21-70 (1988)

[28] Williams R.F., Compositions of contractions , Bol Soc Brasil, Mat 2, 55-59 (1971)

EXPERT SYSTEM OF IMAGE PROCESSING- APPLICATIONS TO THE RECONSTRUCTION OF 3D IMAGES FROM 2D IMAGES

Claude GAUDEAU [1,3], *Vero RAKOTOMALALA* [2], *Ghislaine CARIMALO* [3] *and Jean-François.BERGAMINI* [2]

[1] *CNRS - Laboratoire de Bio-Informatique, 32 rue Emile Zola, 37000 Tours, France*
[2] *Société de Bio-Informatique et de Bio-technologie, 45 rue Emile Zola, 37000 Tours, France*
[3] *CNAM, 2 rue Léo Delibes, 37200 Tours, France*

ABSTRACT

Pattern recognition and 3D reconstruction require the sequential execution of a group of procedures. This necessitates the development of reasoning tools to perform the various operations. First, the defects of a "noisy" image must be filtered. This is accomplished using differentiation and integration operators, which are divided into the categories of context-free, context dependent, global context dependent, and local context dependent.

Other methods include segmentation(isolating specific parts of the image), erosion, and dilatation, The image is then compressed mathematically and "recognized". The expert system uses the method of decision-making graphs to organize the representation and interpretation spaces in order to select the relevant methods and to execute them in the correct order.

RESUME

La reconnaissance de formes et la reconstruction 3D nécessitent l'exécution séquentielle d'un groupe de procédures. Cela exige le développement d'outils de raisonnement afin d'exécuter les diverses opérations. Dans un premier temps, les défauts d'une image doivent être filtrés. Cette filtration est faite en utilisant les opérateurs d'intégration et de différentiation, qui se divisent en contextuels, non contextuels, dépendants du contexte global et du contexte local.

Il existe d'autres méthodes, comme la segmentation (isolation des parties spécifiques de l'image), l'érosion et la dilatation. L'image est compressée mathématiquement et "reconnue".Le rôle du système expert, qui utilise la méthode des graphes décisionnels pour designer ses espaces de représentation et d'interprétation, est de sélectionner les méthodes nécessaires et de les exécuter dans le bon ordre.

1 INTRODUCTION

Pattern recognition and 3D reconstitution of specific objects in imagery require the sequential execution of a group of procedures: differential and integration operators, segmentation, compression, factor analysis and 3D reconstruction.

However, existing softwares include most of the procedures and permit only experienced users to realize an estimated optimum selection based exclusively on their own experience.

It seems essential to offer:

- The potential to realize image processing with no previous knowledge on the part of the inexperienced user,
- For the experienced user, the possibility to memorize his or her own experiment and apply it,
- For all users, the potential to expand their possibilities and enrich the scope of procedures.

This requires the development of reasoning tools to determine careful choices at each step of the treatment, according to the user's goals, the type of images to be processed, the steps already realized and the results obtained.

The sequential working multiplicity pattern requires choices that the experiments can optimize. The expert system turned toward pattern recognition performs a large number of basic operations (erosion, dilation, contour detection, texture analysis, and various filters such as Gradient, Laplacien, Kirsch, etc.)

The expert system allows the operator, after image manipulation (segmentation, recognition of object position, color compression) to reconstruct a 3D picture pattern from a series of 2D pictures taken on unknown incidence and rotation.

The parameters of the 2D pictures are decomposed into a low orthogonal characteristic. On inverse transformation, at least 3 eigen pictures from which the 3D picture will be reconstructed are defined.

2 IMAGE TREATMENTS

This article presents certain methods used to treat a single digitized image usually proceeding from electron microscopy (see Figure 1).

The aim is to isolate and enhance interesting forms by eliminating noises and increasing signals in order to reconstruct the 3D image from 2D images.

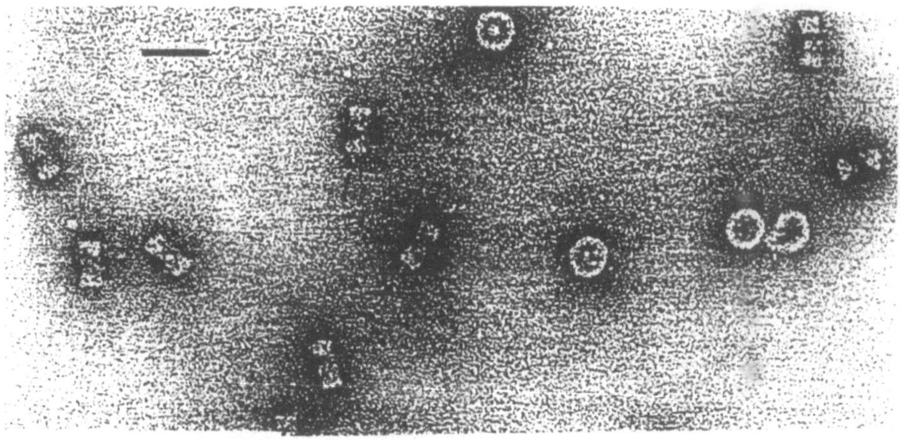

Figure 1. Electron microscopy image of negatively stained immunocomplexes composed of cross-linked hemocyanin and polyclonal fragments specific for functional unit *Ov e*. The bar represents 50nm. From Lambert at al, 1994 in J. Mol. Biol.

2.1 Filtration

A number of defects can appear during image capture: the contrast is insufficient, the image is "*noisy*" because of high frequency noises (scintillations), low frequency noises (stains), the image may also be perturbed by defects due to the material used (deformations by objective lenses, directional noises generated by displacements of the object or of the objective).

These problems must be eliminated or at least diminished in order to isolate the structure to be recognized.

This chapter covers the restoration of images, reduction of noise and signal enhancement .

Segmentation was used to eliminate most of the spurious regions of a picture. With this method and an appropriate parametrization, a picture of human chromosomes was restored [Tanimoto and Pavlidis, 1977]. But as we can see in the next chapter, the segmentation is better at isolating the different patterns of an image.

Noise is usually lowered by differentiation [Gaudeau et al, 1983] and integration operators [Bouhabdallah et al, 1975]. In order to reinforce the necessary information one must make use of differentiation operators, and in order to reduce noise one must make use of integration operators. Optimization is an equilibrium between the two tendencies.

Image transformation by integration operators has provided evidence of molecular transport along the axon microtubules [Travis et al, 1983; Allen et al, 1985]. Differentiation operators, such as Gradient or Laplacien [Pratt, 1978], have been widely applied to biomedical image processing for edge enhancement.

The operators are classified as context-free, context-dependent, global context dependent and local context-dependent (see Figure 2).

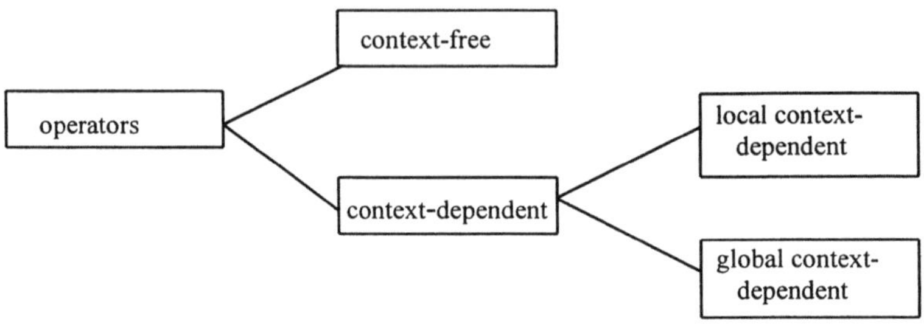

Figure 2. Operator classification

Context-free operators are applied without considering the characteristics of the image, whereas the parameters of the context-dependent operators are chosen in terms of these. For global context-dependent operators global statistical properties are considered (average intensity, grey levels distribution,...) and the comportment of a local context-dependent operator depends on neighborhood information (contour, texture, ...).

Therefore, artificial intelligence is well qualified to choose between the different types of filtering and the parameters involved.

Operators are usually defined as follows [Gaudeau et al., 1990]. M(i,j) is defined as the digitalized picture and R(i,j) as the resultant picture of the treatment. An indexed function is determined on a rectangle as:

$$-n<a<+n \qquad (n,m) \in N$$

$$-m<b<+m$$

We explain the non-contextual operator with Q(a,b) where a and b are invariable and the contextual operator with a' and b' where

$$a' = f(a,M(i,j)) \qquad n = g(M(i,j))$$

$$b' = f(b,M(i,j)) \qquad m=h(M(i,j))$$

in the expression:

$$R(i,j)=\sum_{a=-nb}^{+n}\sum_{b=-m}^{+m} Q(a,b)M(i,j)$$

2.2 Erosion and dilatation

These two transformations are used in the same way as filtration or segmentation. The aim is to increase the signal-noise ratio. Erosion is the search for the minimum in the neighborhood (in binary mode, erosion is defined as AND between values of nearness). Furthermore, dilatation is the search for the maximum in the neighborhood (in binary mode, dilatation is defined as OR between values of nearness).

An appropriate combination of multiple erosions and dilatations involves the enhancement of the signal and the decrease of noise. So forms which must be highlighted become well defined (see Figure 3).

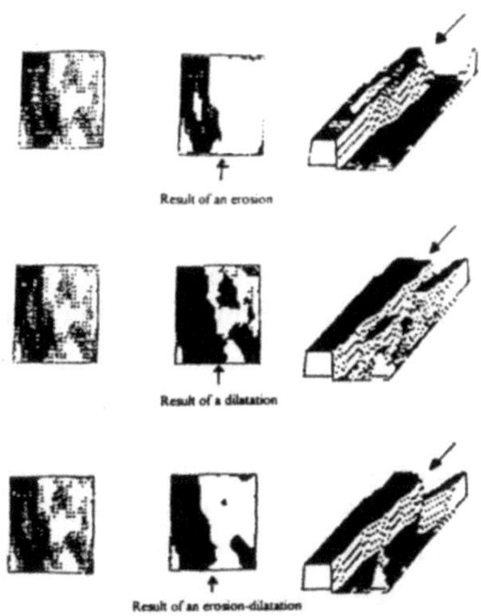

Figure 3. Erosion and dilatation

2.3 Segmentation

Segmentation consists of isolating the image in various areas in which related shapes and textures have to be isolated [Cohen and Feigenbaum, 1981], for example:

- A particular organ in echography,
- Typical molecules in an electron microscopy image, ...

Another use of segmentation consists of the preliminary step of image compression.

A manual segmentation sets a square pattern on an image following vertical and horizontal stapes (grid) defined by the user. Then, vectors of bits are generated corresponding to the windows chosen and capable of being used as entry data in the recognition program. An automatic segmentation can be generated by using orthogonal projections (Figure 4).

The choice of the combination of these transformations is usually the result of human practice. Therefore, this may be considered in writing the rules of an expert system.

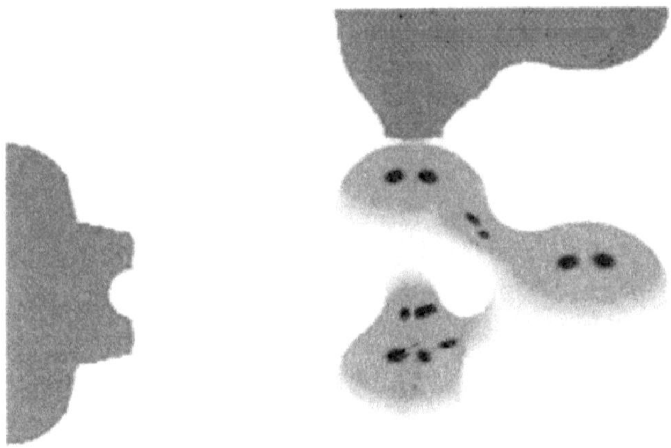

Figure 4. Parameter extraction of form : molecular form signature. Horizontal and vertical histograms permit one to recognize a projection of the molecular form in the micrographs but are not invariant of the form orientation in the plane.

2.4 Image averaging

When the image contains the same object with the same view and the same orientation, averaging can be done. It is commonly used in periodic structures [Saxton and Baumeister, 1 982]. This operation first needs segmentation in order to separate the different objects. A good representation of the general image is shown in Figure 1 and the resulting averaging in Figure 5.

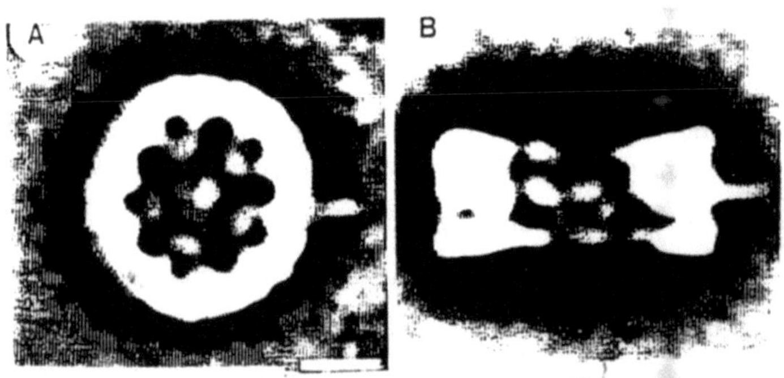

Figure 5. Image average of negatively stained immunocomplexes composed of an anti-*Ov e* fragment and a hemocyanin molecule.From Lambert et al, 1994 in J. Mol. Biol.

2.5 Compression

This step consists in the reduction of initial data from dimensions NxN to (N-k)x(N-k) without losing information. Numerous methods of compression exist [Pratt, 1978; Andrews et al, 1969]. Two are specially considered because they are the best adapted to the problem of form and texture recognition:

- Methods of spatial coding,
- Chatuverdi algorithm.

2.5.1 Methods of spatial coding

Among these methods, the most general is an interpolation in which a pixel plane is fitted by continuous functions. Usually, luminance values are approximated regarding an error set (zero order interpolation).

Electron microscopy pictures of peripheral nerve endings have been treated by this method of compression [Gaudeau and Rossazza, 1 987].

2.5.2 Chatuverdi algorithm

The Chatuverdi algorithm consists of three discrete processing steps. First, after segmentation the fixed image blocks NxN are carried out upon the unitary transform: the cosine transform which allows the image energy to be packed in a few low-ordered transform coefficients (N-k)x(N-k).

In the second step, the even cosine transform is performed on each of the image blocks. The base function of the transform is a class of discrete Chebyshef polynomials.

Finally, the two-dimensional transform matrix is arranged in one-dimensional array by reading the matrix along the cross diagonal.

Each image block is coded, after transformation by a vector.

The choice of a method is greatly influenced by the characteristics of the image. Here too, artificial intelligence plays a decisive role.

3 Pattern Recognition and Correspondence Analysis

Two methods are available to make computers recognize forms or structures:

- Factor and correspondence analysis [van Heel, 1984]
- Classification [Gaudeau, 1975]

3.1 Parameter extraction of forms: form signatures

The aim of this method is to reduce the number of images by classifying them in families constituting the same view of the object with different orientations. The center of gravity G of each form is calculated and concentric crowns centered on G are computed. The histograms of the different crowns are calculated (see Figures 5 and 6). The pictures which have the same crown histograms are classified in K families. The correspondence matrix thus becomes an N/K x N/K.

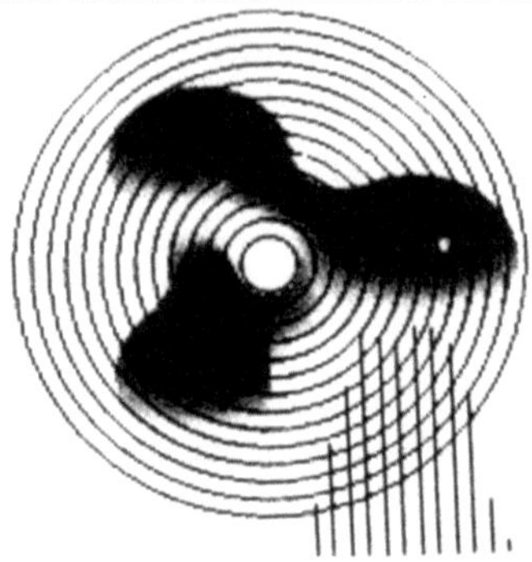

Figure 6. Parameter extraction of form : molecular form signature. Crown histograms are invariant of the form orientation in a plane.

3.2 Factor and correspondence analysis

Correspondence analysis has been used in molecular analysis [van Heel and Frank, 198 1] and, recently, in the study of machine vision [Pentland et Al., 1 994]. If the Chatuverdi algorithm is applied, we have to deal with vectors, so factor analysis is used, whereas for images, correspondence analysis is used. The method is explained considering only correspondence analysis. For N pictures, a NxN correspondence matrix is constructed. Each picture is thus represented by a linear combinaison of eigenvectors (see Figure 7).

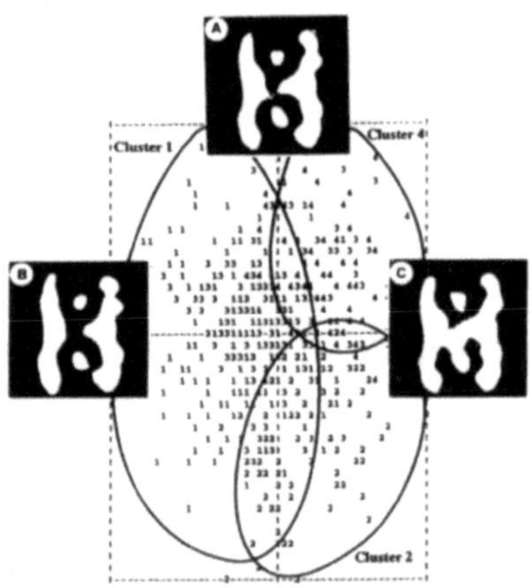

Figure 7. Separation of a (1/1) mixture α_2 M-MA molecules nonconjugated and conjugated to cytochrome c by correspondence analysis. The inserts corresponds to (A) the average map of the whole image set and the average maps of the images with (B) negative and (C) positive values of the first factorial axis. From Lambert et al, 1994 in J. Struct. Biol.

3.3 Classification

This method involves a treatment of the vectors attributed to each image block at the end of the compression step. The collection of vectors is confronted with a set of classes proposed by the operator according to the type of pictures.

$P=(P_l,...,P_m),P_i \varepsilon R_m$ defines the collection of vectors,

$C=(c_l,...,c_{m-k})$ is the set of classes.

After standardization, a random choice of a vector is made among the set of vectors and compared with the classes by a discrimination operator requiring notions of angle, distance and area. All vectors are attributed to classes for which they obtain the best affinity. This method has produced good results in distinguishing molecular textures in electronic microscopy pictures [Pentland et Al., 1994].

3.4 3D model construction

Three steps are implicated in the process of constructing 3D structures. First, pattern subunits must be coded or compressed so as to facilitate spatial correlations. Then, correlation correspondence matrices, eigenvectors and eigenvalues are computed and finally from at least three eigenvectors, the 3D object will be reconstructed. If a transformation was made (Chatuverdi for example), the inverse transformation is necessary.

3.4.1 Freeman coding

The Freeman code is used in the reconstruction method. It consists of coding from subunits into a chain of numbers representing line segments connecting adjacent pixels in a horizontal, vertical or diagonal direction [Gaudeau et al, 1988]. This coding method does not necessarily seek to produce an efficient code in the sense of minimizing the number of bits required to describe a pattern boundary, but rather provides some manipulative properties.

3.4.2 Correlation

Correlation methods involve two blocks coded by Freeman codes or another compression. The first Freeman chain is read clockwise and the second counterclockwise. A correlation will appear between the two numerical chains only if a relation of congruence module 4 exists between two elements of the same order within the two chains to be compared.

4 EXPERT SYSTEM

As we mentioned earlier, artificial intelligence must play a role in each determinant step of the treatment involving the basis of knowledge adapted for the different operations.

The sequential working multiplicity is governed by expertise which acts at nearly all levels of the treatment. It performs a large number of basic operations (erosion, dilatation, contour detection, ...) [Gaudeau et al., 1990]. The system allows the operator, after image manipulation, to reconstruct a 3D picture pattern from a series of 2D pictures of unknown incidence and rotation. Texture analysis can be used to optimize 3D reconstruction.

4.1 Interpretation, representation and application spaces

The expert system philosophy works in the three spaces called interpretation space, representation space and application space (see Figure 8). In the interpretation space, reasoning starts from a concept and deduces properties of characteristics which are grouped in a collective class [Pentland et Al., 1994].

Elsewhere, reasoning starts from a concrete object and, by induction, structures a representation space with morphological characteristics according to the object. This induction can entail a processed risk of few indicative characteristics toward the concept. The expert system's purpose is to reduce during the treatment the representation space dimensions. The application space is composed of applications from the representation space into the interpretation space.

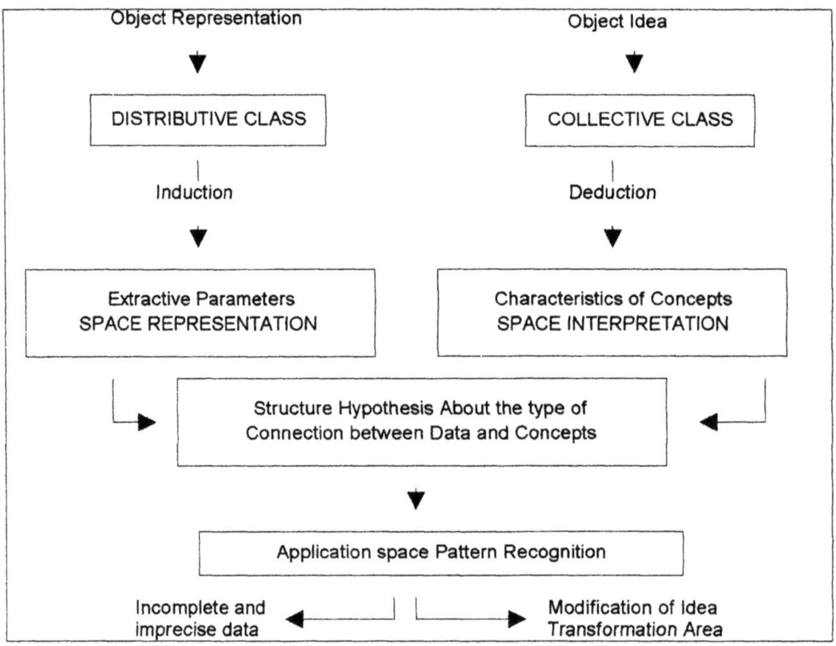

Figure 8. Representation of pattern recognition reasoning.

The classification, which is an important phase of the recognition process, starts from the collective class and divides it in order to construct a distributive class. A subsequent operation of reassembly composes new collective classes.

4.2 Organization of the expert system

An expert system is conceived in order to structure representation and interpretation spaces using successive induction and deduction operations. Basic structure of essential knowledge, useful in the constitution of the two preceding spaces, is over-simplified by decision-making graphs (see Figure 9): Gi(Xi,Yi) with i = (l, 2).

- For the interpretation space,

Xl is a set of nodes where every node corresponds to questions progressively helping to define the operator's aim and Yl is a set of arcs where every arc corresponds to the operator's answer. The result of this graph leads to parametrisation of G2.

- For the interpretation space,

X2 is a set of nodes where every node corresponds to a procedure. Y2 is a set of arcs where every arc corresponds to a decision between two procedures according to the results of a procedure of the preceding graphs Gl and G2.

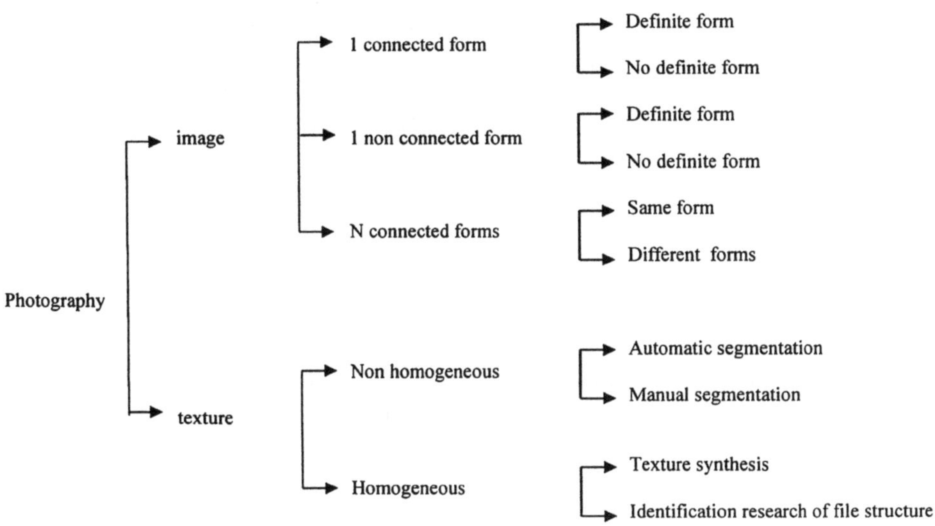

Figure 9. Decision-making graph.

The basis of knowledge of the expert system is comprised of development rules in the decision-making graphs Gl and G2.

In conclusion, Gl and G2 are graphs of two automatons interconnected in series in which the outputs of Gl partly make up the inputs of G2.

Examples of rules and of consultation of the expert system are shown respectively in Figures l0 and 1 l .

Rule number 1:
if image: True
then texture: True
Rule number 2:
if Texture: True
if Homogeneous image: True
then Segmentation: True
Rule number 3:
if Texture: True
if Homogeneous image: True
then Filtering: True
Rule number 4:
if image: False
then Connection: True
Rule number 5:
if Connection: True
if A single connected or non-connected image: True
the Standard Form: True
Rule number 6:
if Connection: True
if A single connected or non-connected image: False
then Comparison of forms: True
Rule number 7:
if Comparison of forms: True
if Identical forms N: True
then N forms to define: True

Figure 10. Examples of expert system rules

Enter the text of the fact (or ESC):
Image
True or False? T/F: t

Enter the text of the fact (or ESC):
Homogeneous image
True or False? T/F: t
1 deduce filtering true

Enter the text of the fact (or ESC):
Noise present image
True or False? T/F: f
I deduce Gauss' filtering
followed by binarisation true

Enter the text of the fact (or ESC):
Image
True or False? T/F: f
One single connected form or not
True or False? T/F: t

Enter the text of the fact (or ESC):
Definite form
True or False? T/F: t
I deduct contour analysis

Enter the text of the fact (or ESC):
Clear contours
True or False? T/F: f
I deduct erosion true

Figure 11. Example of expert system consultation by prechaining

CONCLUSION

A large panel of tools is now available. Images can be restored by filtering using operators. Then segmentation permits the isolation of relevant forms in the pictures. Erosion and dilatation combinations can enhance signals to obtain pictures of good quality. If numerous images are available, compression algorithms are recommended to reduce CPU time and to free the memory.

After these transformations, the forms corresponding to the different views of the object are well defined and pattern recognition is thus carried out by classification and correspondence analysis.

Finally, as an eigenvector is attributed to each form, 3D picture reconstruction is realized using Freeman coding and the correlation method.

The role of the expert system is to govern the whole treatment. It must make the best choices between the different tools and conduct them in the correct order.

Acknowledgments

We would like to thank for their collaboration:
M. L. Chu, M. G. Dauphin, M. J.P Francoz, Mme S. Benoist, M. T.Coue.

Keywords

Image processing, pattern recognition, 3D image reconstruction, expert system, electron microscopy

Reference

[1] Allen R.P., Weiss D.G., Hayden J.H., Brown T.D., Fujiwake H., Simpson M., Gliding of movement of end bidimensional transport along single native microtubules from Squid axoplasms: evidence for an active role of microtubules in cytoplasmic transport, J.Cell.Biology, 100(5), 1736-1752 (1985).

[2] Andrews H.C., Kane J., Pratt W.K., Hadamard transform image coding, Proc.IEEE, 57(1), 58-68 (1969).

[3] Boisset N., Radermacher M., Grassucci R., Taveau J-C., Liu W., Lamy J., Frank J., Lamy, J.N., Three dimensional immunoelectron microscopy of scorpion hemocyanin labeled with amonoclonal fab fragment, J.Struct.Biol.,111, 234-244 (1993).

[4] Bouhabdallah M., Gaudeau C., Kurc C., Longevialle C., Mallet C., Morand P., Thouvenot J., Automatic analysis of cineangiorcardiography, Computer in Cardiology, IEEE, (1975), 267.

[5] Cohen, Feigenbaum, The handbook of Intelligence artificial, Heuristich Press, vol III (1981).

[6] Frank J., Liu W., Boisset N., Classification and 3D variance estimation: Complementary tools in the 3D reconstruction of the macromolecules, Proc. 10th Eur.Cong.Electron Microsc.,1, 427- 429 (1992).

[7] Gaudeau C., Thèse de biologie humaine sur la reconnaissance des états d'un système physiologique et ses processus aléatoires, Identification par des modèles stochastiques,Tours (1975).

[8] Gaudeau C., Rossazza J-P., Système de traitement d'images par microordinateur: Application de l 'analyse factorielle des correspondances à la reconstruction d'images 3D à partir d'images 2D prises sous des incidences inconnues, J.Microsc.Spectrosc.Electron.,12, 155-158 (1987).

[9] Gaudeau C., Bosma M., Rossazza J-P., Spatial modelling and simulation of organized macromolecular structure by means of factor analysis of electron microscopy pictures, Mol. Phys. Chem. Biol., 1, 129-157 (1988).

[10] Gaudeau C., Rossazza J-P., Normand, B., Gouthiere, L., Cassis, N., Application des systèmes experts en imagerie, 115e Congr.Nat.Soc.Savantes, 305-322 (1990).

[11] Gaudeau, C., Routier, G., Diedzina, D., Longevialle, C., Image sequence processing and pattern recognition of bio-medical picture. Image sequence processing and dynamic scene analysis, 2 Springer-Verlag, Berlin, 625 (1983).

[12] Laforge J.M., Classes méréologiques et construction logique. Doctorat de logique mathématique, Gand (1973).

[13] Lambert, O., Boisset, N., Penczek, P., Lamy, J., Taveau, J-C., Frank, J., Lamy, J.N., An approach to the intramolecular localization of the thiol ester bonds in internal cavity of human a2-Macroglobulin based on correspondence analysis, J.Mol.Biol ,238(1), 75-87 (1994).

[14] Lambert , O., Boisset, N., Pochon, F., Delain, E., Lamy, J.N., Quaternary structure of OctopusVulygaris Hemocyanin. 3D reconstruction from frozen hydrated specimens and intramolecular location of functional units O_ve and O_Vb, J.Struct.Biol,112, 148-159 (1994).

[15] Liu W., 3D variance of weighted back-projection reconstruction and its applications to the detection of 3D particles conformational changes Proc. 49th Ann.Meet.Electron Microsc.Soc.Am., 1, 542-543 (1991).

[16] Penczek P., Radermacher M., Frank J., Three dimensional reconstruction of single particles embedded in ice, Ultramicroscopy, 40, 33-53 (1992).

[17] Pentland A., Trevor D., Azarbayejani A., Sclaroff S., Towards machine vision in complex environments, 9ème Congrés de Reconnaissance des Formes et Intelligence Artificielle (1994).

[18] Pratt, W.K., Digital Image Processing, John Wiley, New York (1978).

[19] Rademacher, M., Three dimensional reconstruction of single particles of random and nonrandom tile series, J.Electron.Microsc.Tech., 2, 359-394 (1988).

[20] Radermacher, M., Srivastava, S., Frank, J., The structure of the 50S ribosomal subunit from E.Coli in frozen hydrated preparation reconstructed with SECReT, Proc. 10th Eur.Cong.Electron Microsc., 3, 19-20 (1992).

[21] Saxton W.O., Baumeister, W., The correlation averaging of a regularly arranged bacterial cell envelope protein, J.Microsc., 127(2), 127-138 (1982).

[22] Tanimoto S.L., Pavlidis T., The editing of picture segmentation using local analysis of graph, Comm.ACM, 20(4), 223-229 (1977).

[23] Travis J., Kenealy J., Allen R.P., Studies on the motivity offoramifera tomell. The dynamic microtubular skeleton of the reticulopodial network of allogromia laticolaris, J.Cell.Biology, 97(6), 1668-1676 (1983).

[24] van Heel M., Multivariate statistical classification of noisy images (randomly oriented biological macromolecules), Ultramicroscopy, 13, 165-184 (1984).

[25] van Heel M., Franck J., Use of multivariate statistics in analysing the images of biological macromolecules, Ultramicroscopy, 6, 187- 194 (1981).

Chapter 6

SYMMETRY CONCEPTS AND APPLICATIONS

DEDUCING MOLECULAR SYMMETRY BY TOPOCHROMATIC PERMUTATIONS

Michel LOUKIANOFF, Aliette COSSE-BARBI and Jacques-Emile DUBOIS

ITODYS, Université Denis Diderot, 1 rue Guy de la Brosse, 75005 Paris, France
E-mail: cosse@paris7.jussieu.fr

ABSTRACT

A new method for elucidating the symmetry of rigid molecular objects is described. In contrast to current methods for determining symmetry point groups (SPG), which are based on geometrical data, this method is based on topochromatic permutations, and compares the sub-graphs of permuted elements with the invariant sub-graphs. The correspondence between a topochromatic permutation and a geometrical symmetry operation is achieved by means of a « permutational chart », derived in part from the usual « Chemical Flow Chart ». Optimization of algorithms related to matrix permutation improves the efficiency of the method.

RESUME

Une nouvelle méthode d'élucidation de la symétrie d'objets rigides moléculaires est décrite. Contrairement aux méthodes courantes de détermination des groupes ponctuels de symétrie (SPG) qui privilégient l'utilisation des données géométriques, notre méthode s'appuie sur les permutations topochromatiques et confronte les sous-graphes d'éléments permutés aux sous-graphes d'éléments invariants. La correspondance entre une permutation topochromatique et une opération de symétrie géométrique est effectuée à l'aide d'un "organigramme de permutations" en partie dérivé de l'organigramme "Chemical Flow Chart" habituellement utilisé. L'efficacité de la méthode est accentuée grâce à l'optimisation des opérations liées aux permutations matricielles.

1 INTRODUCTION

Normally, the determination of the symmetry point group (SPG), or of more refined symmetry descriptions (framework group [(1)], Fujita notation [(2)]), are based mainly on geometrical data. These can be incomplete, whether they come from a 3-D molecular model or a 2-D projection and analogy structures of known symmetry.

This symmetry groups determination method is difficult to apply, as it requires a good perception of spatial relationships and an excellent knowledge of group theory.

One of the essential objectives of our work on the visualization and the representation of molecular structures is to help the non-expert work out molecular symmetries. To do this, we start from the observation that in principle a given topology and chromatism (connectivity and nature of the atoms) do not allow any geometry : nonetheless a small number of geometries are shown to be compatible with a given molecular graph. It is then possible to reduce the compatible symmetries by using topological concepts and permutations which retain the molecular graph.

On the basis of certain relationships between the topological and geometrical fields, we propose a procedure for counting the equivalent permutations arising from a simple extended formula satisfying the site connectivities. This leads to a set of very numerous topochromatic permutations (the topochromatic permutation group, TPG) as do the symmetry supergroups for non-rigid structures[(3)]. This TPG leads to the SPG by means of a geometric interpretation of the topochromatic permutations. The procedure eliminates most of the purely local symmetries and ignores permutations which have no geometric equivalent. Next, a permutational flow chart selects sets of possible geometric solutions, according to the notion of group.

2 RELATIONSHIP BETWEEN TOPOLOGICAL AND GEOMETRIC FIELDS

The relationship between these fields is not reciprocal : if we know the nature and the positions of the atoms of a molecule (chromatism and geometry) we can deduce its graph (topology). On the other hand, a given chromatic graph can express several molecular geometries. We need first, therefore, to discuss certain aspects of the relationships between topological and geometric fields.

2.1 From the geometric to the topochromatic field

Any symmetry point operation (inversion, reflection, proper or improper rotation) on a molecular structure exchanges the positions of the atoms in the geometric field. This exchange corresponds to a permutation of the atoms and their bonds in the molecular graph (permutation of equivalent sub-graphs, referred to as « topochromatic »). This does not modify the molecular graph itself: since the bonds permute with the atoms, the connections are always retained[(4)].

The set of these permutations, associated with an internal composition rule, forms a topochromatic symmetry group (TSG), whereas the 3D symmetry operations on molecular structure constitute a symmetry point group (SPG).

The atoms located on an invariable element (point, axis or plane) of a 3D symmetry operation are necessarily the only ones which are not affected by the associated topochromatic permutation.

2.2 From the topochromatic to the geometric field

If the topology is taken as the starting point, one can observe permutations of atoms (and their bonds) which retain the topology and the chromatism of the molecular graph but correspond to no 3D symmetry operation. This is the case when the atoms which are unaffected by a topochromatic permutation cannot be placed on a same 3D symmetry element. Deducing a molecular geometry from a topological situation requires therefore that the topochromatic permutations be screened, only those which have a relevant geometric interpretation being kept. Our approach consists therefore in generating the topochromatic permutations in order to search for the corresponding symmetry operations with the help of certain rules of thumb.

3 DETERMINATION OF TOPOCHROMATIC PERMUTATIONS

The permutations which retain a chromatic graph exchange only the nodes of the same type and with the same overall environments. In fact, permuting the nodes and the edges of a graph without modifying the topology amounts to simply exchanging the names of the nodes in such a way that their neighbors (connected nodes) are retained (by their names) after the permutation. This is an overall modification of the labelling (or numbering) of the nodes which in no way affects their relative topological positions. The equivalent labels are determined on the basis of a canonical numbering of all the atoms of the molecular structure[(5)]. The most frequently used numbering procedures are those of Morgan/CAS and DUPEL/DARC. In this application we have chosen to base our procedure on an adjacency matrix (AM) which, by definition, is directly related to the numbering chosen. Two topologically equivalent numbering systems are recognized by the fact that their adjacency matrices are identical.

It should be remembered that all the equivalent permutations or labellings are required in order to deduce the symmetry data sought. However, it would be tedious to generate all the possible permutations in order to compare their AMs. We have therefore devised a procedure for optimizing this operation. We are able to reject a large set of irrelevant permutations by using *(i)* reduced atom equivalence classes, *(ii)* a purely concentric scan of the graph and *(iii)* an « optimal matrix » which undergoes very few modifications during the AM search procedure.

The example of the described procedure illustrated in Figure 1 (table A2) demonstrates the choice of such a matrix and shows also the generation of the permutations selected (p_1, p_2, p_3, p_4, p_5, p_6, p_7 and p_8). The algorithm constructed for this work makes it possible to handle even very symmetrical structures rapidly, thanks to economic management of the recurrent parameters and the way in which the degree of recurrence never exceeds the number of atoms in the structure. For the C60 fullerene the calculation is ten times faster by this method (3 s for a 386 SX computer at 33 Mhz) than by the recent procedure of Bohanec & Perdih[6] despite the pretreatment used by these authors.

4 LISTING OF CANDIDATE SYMMETRY OPERATIONS BY TOPOCHROMATIC PERMUTATION

Every 3D symmetry operation is defined by its degree, its type (proper or improper rotation), and its invariable geometric element (or "symmetry element"). In a molecular structure, the set of variable atoms in a symmetry operation is identical with the set of atoms exchanged by the associated topochromatic permutation (the invariable atoms correspond in the same way). We propose therefore to deduce the degree of a candidate operation expressing a topochromatic permutation by observing the behavior of the permuted atoms, while the symmetry element will be suggested by a non-permuted sub-graph. However, the type (proper or improper) of a 3D operation is a purely geometric piece of information indicating the way in which variable objects are redistributed in space around invariable objects; in other words, the type has no topochromatic meaning. The type of the selected symmetry operation should therefore simply ensure that the chosen degree and symmetry element are compatible.

The candidate 3D operations are selected under the following constraints:

i) The degree of a symmetry operation is identical with that of the topochromatic permutation studied (it is the number of times that this operation must be applied to obtain an identity operation). Since a topochromatic permutation is always circular, its degree (or order) is the smallest common factor of the sizes (number of nodes) of the elementary permutations of which it is composed.
Example : p=(abcdef)(ghij), degree(p)=SCF(6,4)=12.

ii) The two types of possible operations are not always compatible with the degree chosen.
A degree 1 permutation always expresses an identity operation, but sometimes also an infinite rotation or a σ mirror symmetry acting through the molecular plane.

A degree 2 permutation can correspond to a proper rotation c_2, a point inversion i or a mirror symmetry σ.
A permutation with an odd degree (d>3) always expresses a proper c_d rotation.
A permutation with an even degree (d>2) can express an improper rotation s_d or a proper rotation c_d, or an improper rotation $s_{d/2}$ when the degree d is not a multiple of four.

iii) The symmetry element must be compatible with the type of operation (i, σ, c_d or s_d) and with the invariable sub-graph. One must therefore check that the topochromatic graph which is invariable to the permutation studied (and only to this one) is compatible with the invariable geometric element of the proposed symmetry operation.

The example of the procedure shows how the candidate symmetry operations are rejected by observing that the topochromatic sub-graph of the invariables does not fit any planar, linear or point geometric representation. This is the situation for operation s_4 which is a candidate for permutation p_6, and the purely local or geometrically insignificant symmetry operations associated with permutation (p_2 and p_3) and (p_6 and p_7), respectively. We have coded a set of topochromatic relationships which, for each permutation isolated in the first instance, selects one or several candidate symmetries. It remains then to choose inter-list combinations of these operations corresponding to a meaningful geometric description.

5 INTER-LIST COMBINATIONS OF CANDIDATE SYMMETRY OPERATIONS FOR COHERENT SOLUTIONS

The candidate symmetry operations cannot be considered independently of each other, since the set of these operations is a group (the SPG). We shall therefore search for the topochromatic permutations which satisfy a series of hypotheses concerning the presence of operations which make up the point sub-groups. To do this, we use a « permutational flow chart » which is derived from the « chemical flow chart » used by chemists for the visual determination of symmetry point groups. The two charts are in fact based on the idea of the symmetry sub-group, but differ markedly in their goals and deductive procedures :

- The chemist starts with the geometry (at least a rough overall image of it) and deduces which point group the structure belongs to by means of a binary chart (such as « yes, there is a c_3 rotation axis » or « no, there is no c_3 axis »). It should be noted that the point group is independent of the structure which belongs to it; it defines a symmetry class.
- Our approach is different. As if we were blind, we have no geometrical support (we shall refine this point later on). Rather than a point group, we consider candidate symmetry operations and the behaviour of the atoms in the structure subjected to these operations. The candidate operations are plausible geometric interpretations (there are generally several) for the topochromatic permutations (Figure 2).Thus, our chart is not organized in terms of « yes/no » but rather in terms of impossibilities (such as « there can be no c_3 rotation axis perpendicular to

the mirror plane previously defined » or, more generally, « this geometric interpretation is not consistent with the other proposal advanced for other permu-

Figure 1. Example of the procedure

Determination of the topochromatic permutations :

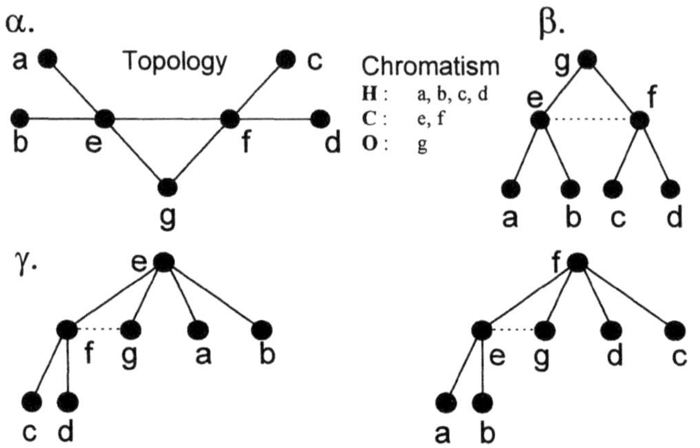

A1	g	e	f	a	b	c	d
g	0	1	1	.	.	.	.
e	1	0	.	1	1	.	.
f	1	.	0	.	.	1	1
a	.	1	.	0	.	.	.
b	.	1	.	.	0	.	.
c	.	.	1	.	.	0	.
d	.	.	1	.	.	.	0

A2	e	f	g	a	b	c	d
e	0	1	1	1	1	.	.
f	1	0	1	.	.	1	1
g	1	1	0	.	.	.	.
a	1	.	.	0	.	.	.
b	1	.	.	.	0	.	.
c	.	1	.	.	.	0	.
d	.	1	.	.	.	.	0

δ.

	1	2	3	4	5	6	7	permutations	degrees	invariable atoms
p_1	e	f	g	a	b	c	d	()	1	a, b, c, d, e, f, g
p_2	e	f	g	a	b	d	c	(c d)	2	a, b, e, f, g
p_3	e	f	g	b	a	c	d	(a b)	2	c, d, e, f, g
p_4	e	f	g	b	a	d	c	(a b) (c d)	2	e, f, g
p_5	f	e	g	c	d	a	b	(a c) (b d) (e f)	2	g
p_6	f	e	g	c	d	b	a	(a c b d) (e f)	4	g
p_7	f	e	g	d	c	a	b	(a d b c) (e f)	4	g
p_8	f	e	g	d	c	b	a	(a d) (b c) (e f)	2	g

In β : A « breadthwise first » scanning of the graph (α) and its adjacency matrix.
In γ : Two equivalent « breadthwise first » scannings of the graph (α) and their adjacency matrices.
In δ : Lists of the equivalent atom permutations (p_1, p_2, ..., p_8) associated with the A2 matrix.

Lists of candidate symmetry operations for each topochromatic permutation :

p_1 : All the atoms are invariable but they are not located on the same plane because of the presence of tetracoordinate carbon atoms; the only operation which fits this permutation is the **identity operation (E)**.

p_2 : Atoms *a, b, e* and *g* are invariable to this permutation. However, atom *e* is a carbon whose neighbors cannot be located on the same plane. This is a **local** permutation. It is not considered in the following treatments.

p_3 : **Local** permutation (see comments under p_2).

p_4 : Atoms *e, f* and *g* are invariables and belong to the same ring; they are therefore on a plane of symmetry. This permutation must correspond to a **mirror symmetry** σ.

p_5 : The fact that the degree is 2 suggests three possibilities, but inversion is impossible since the invariable atom *g* cannot be located between the symmetrical atoms *e* and *f* (ring). The remaining possibilities are a **proper rotation** c_2 or a **mirror symmetry** σ.

p_6 : This permutation cannot correspond to a 4th degree proper rotation, since it contains an elementary permutation (*e f*) of degree 2. Moreover, the improper rotation s_4 does not fit either, as this would require that the invariable atom *g* be placed half-way between *e* and *f* on the improper rotation axis, which is not possible because (*e, f, g*) is a ring. This permutation has therefore **no corresponding 3D operation**.

p_7 : **No corresponding 3D operation**.

p_8 : For the same reasons as for p_5, this permutation corresponds to a **proper rotation** c_2 or a **mirror symmetry** σ.

Figure 2. Inter-list organization of the candidate symmetry operations for coherent solutions.

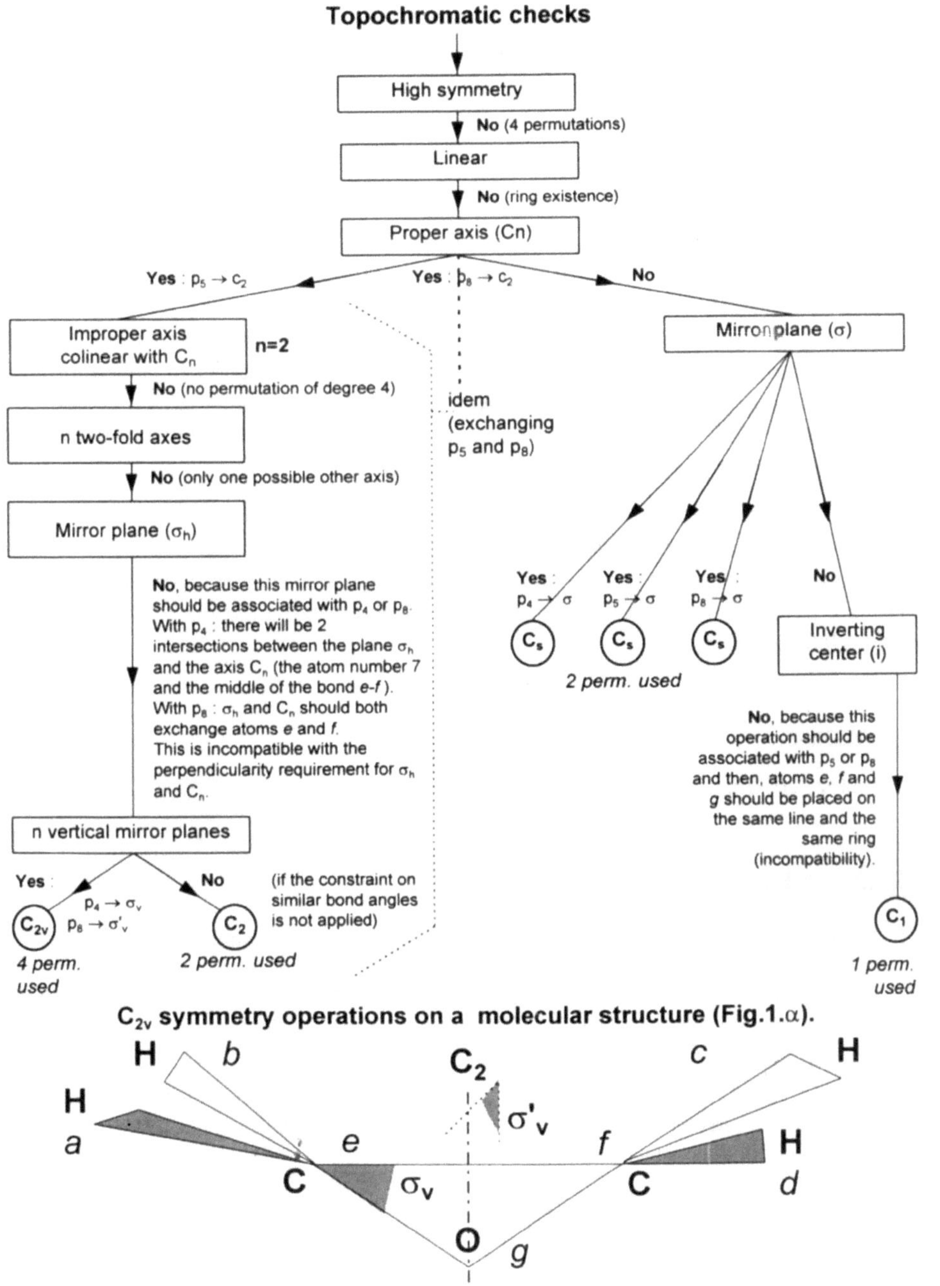

-tations ») or plausible hypotheses (« yes, there can be a c_3 axis », etc.). In this last case several solutions sometimes fit the same hypothesis, since replying « yes » to a hypothesis amounts to choosing the symmetry operations and thus selecting one of the sub-groups of permutations which interpret a geometric invariable element. Finally, if several solutions are found (sets of plausible operations) the symmetry point groups which they define can, nevertheless, be different.

6 CONCLUSION

We have proposed a method for elucidating molecular symmetries based on the topological field. This is not a blind systematic procedure. In fact, we have various guides and reference points in the geometric space. The most obvious of these is that it is forbidden to superimpose two atoms in the geometric space. In the same way, two bonds never intersect. Various knowledge levels (0...n) need to be considered. The example which we give here (Fig.1 and Fig.2) describes the entire procedure at level 1 of standardization.

It is important that the knowledge or constraint level of standardization be perfectly defined before searching for symmetry : if the constraints are too rigid no solution may be found, whereas the lowest level (the topological level) can generate too many candidate symmetries some of which would be unrealistic. In any case, the choice must be left to the operator.

KEYWORDS

Symmetry, Topology, Permutation, Ordering.

REFERENCES

[1] Pople J.A., Classification of Molecular Symmetry by Framework Groups, J. Am. Chem. Soc., 102, 4615-4622 (1980).

[2] Fujita S.F., Systematic Classification of Molecular Symmetries, Symmetry and Enumeration in Chemistry, Springer-Verlag, 63-73 (1991).

[3] Longuet-Higgins H.C., The Symmetry Groups of Non-Rigid Molecules, Molecular Physics, 6, 445-460 (1963).

[4] Loukianoff M., Cossé-Barbi A., Dubois J.-E., From Topochromatical Permutations to Molecular Symmetry Operations, C.R. Acad. Sci. Paris, t.318, Série II, 8, 1043-1049 (1994).

[5] Razinger M., Balasubramanian K., Munk M.E., Graph Automorphism Perception Algorithms in Computer-Enhanced Structure Elucidation, J. Chem. Inf. Comput. Sci, 33, 197-201 (1993).

[6] Bohanec S., Perdih, M., Symmetry of Chemical Structures : A Novel Method of Graph Automorphism Group Determination, J. Chem. Inf. Comput. Sci, 33, 719-726 (1993).

CHIRBASE : A DATABASE UTILIZING A MOLECULAR RECOGNITION APPROACH FOR SELECTION OF A CHIRAL CHROMATOGRAPHIC SYSTEM.

Christian ROUSSEL and Patrick. PIRAS

ENSSPICAM, Av. Escadrille Normandie-Niemen, F13397 Marseille Cedex, France

ABSTRACT

The direct resolution of enantiomers by high-performance liquid chromatography using chiral stationary phases is today one of the most powerful techniques employed in pharmaceutical, agrochemical, asymmetric synthesis and related fields. The CHIRBASE *database was initially developed in 1988 on Chembase®. It now contains more than 30,000 separations of 10,000 compounds and has been proved to be a very efficient tool. An interesting way of extending the current system is by the implementation of* CHIRBASE *into* REACCS™ *and* ISIS™/Host *systems (minicomputer or UNIX workstation). ISIS adds facilities in the searching procedures and provides a large range of chemical descriptors which can be used in similarity searching, and in structure activity studies. The database technique which is being developed links 3D searching techniques and molecular modeling.*

RESUME

Les techniques de séparation d'énantiomères par chromatographie liquide sur phases stationnaires chirales (PSC) connaissent depuis cette dernière décennie un développement considérable. Nous décrivons ici les résultats d'un programme de développement qui a tout d'abord conduit à la construction de la base de données moléculaires CHIRBASE *développée sous* ChemBase®. *Nous poursuivons le développement de* CHIRBASE *sur une architecture logicielle* (REACCS™ *et* ISIS™/Host) *et matérielle (VAX) qui apporte des fonctions nouvelles utiles pour développer un système expert intégrant une base de données 3D et un système de modélisation moléculaire.*

1 INTRODUCTION

Enantiomers present identical physical and chemical properties; however in the presence of chiral frameworks the identity of the properties does not hold anymore.

The resolution of enantiomers by high-performance liquid chromatography (HPLC) on chiral stationary phases (CSPs) is based on the formation of labile diastereoisomeric complexes between the chiral selector and the enantiomers. During the analysis, the internal energies of the two diastereomers are not identical giving rise to separation. The enantiomer having the larger interaction energy with the CSP is more retained than the other one which is eluted first. It is important to note that chiral separation by HPLC results from molecular interactions between the enantiomers and the CSP. These molecular interactions involved in the process of the formation of the solute-CSP complexes can be used as criteria for the division of the CSPs into two classes (1,2) :

- Molecular CSPs where simple chiral compounds are bonded to an achiral matrix. It is assumed here that chiral recognition requires at least three simultaneous interactions between the partners involved. The phases developed by Pirkle are representative of that group (3).
- Supramolecular CSPs which are represented by synthetic and natural polymers such as proteins or polysaccharides. Despite the rather broad applications of these CSP's, the molecular recognition mechanisms are not well understood and therefore the prediction of the best CSP fitting a given analysis remains empirical.

The unambigous determination of enantiomeric excess is an important task in the research concerned with pharmaceutical, agrochemical and related fields because of growing regulatory concern. As many chiral HPLC-CSP systems have been described for the resolution of racemic compounds into enantiomers (4,5,6) it is thus not easy to find a suitable system for a particular separation problem.

An attempt to address this problem is the use of computer-assisted molecular modeling (7,8). This approach should allow the elution order prediction or CSP selection on the basis of the comparison of the energies of different association complexes generated by the computer. One of the problems which can arise is the number of association complexes which make the comparison time consuming and not easily achievable because the energy differences are often too small. A second problem is the solvent effects which are not taken into account.

2 The CHIRBASE Project

The method proposed by *CHIRBASE* is also a computational approach but it involves structural feature searches in a knowledge base of information which contains both successful and unsuccessful chiral separations. In our approach we assume that if similar key features are present in two compounds, their enantiomers could form similar diastereoisomeric complexes and be resolved with the same CSP. This implies that a structure-based retrieval system must be able to search on enantiomers and/or CSPs.

The database *CHIRBASE* established at the University of Aix-Marseille was first built on a PC-Based-software ChemBase® (9,10). The size of *CHIRBASE* is increasing rapidly: there are at present 30 000 separations using around 700 different CSPs. Over 30 fields including structures of enantiomers and CSPs are searchable. In three years, the scope of *CHIRBASE* has been extended to the gas chromatography and thin layer methods. These are developed by experts in the field at the University of Tübingen in Germany and will be merged in the future with the liquid chromatography database.

Searching the database has already been described (11,12). It can be carried out in four stages going from specific to generic queries: exact structure search, substructure search, search of fragments and search of compound of the same family. The search of fragments is particularly relevant for new compounds. The target compound is cut into four generic fragments around the asymmetric carbon. These generic fragments reflect the molecular properties around the chiral carbon. A substructure search is then performed to find in *CHIRBASE* the molecules which have a similar chemical environment. The ChemBase version of *CHIRBASE* achieves many of its original objectives by providing solutions from structural analogue searches in a database of 2D structures. However, if the target molecule is small, there may be a very large number of hits after a substructure search in which case it is necessary to reduce the list to a manageable number by the use of some additional criteria. In addition, the database coding and facilities of ChemBase do not enable further developments with respect to our research goals which focus on :

• Finding methods that can be used to select and prioritize a CSP when nothing is known about the structural requirements for achieving a separation.
• Rationalizing and predicting activity at the CSP level.

3 New Developments

An obvious way of extending the current system is by implementing *CHIRBASE* into REACCS™ and MACCS™ systems which can now be accessed via a distributed information system based on a client/server architecture marketed by MDL: ISIS™/Base (PC Windows, Macintosh) and ISIS™/Host (minicomputer or UNIX workstation). ISIS adds facilities in the fragmentation procedure and provides a large range of chemical descriptors which can be used in substructure and similarity searches, and in structure activity studies. ISIS™/Base provides the user with an easy-to-use query formulation interface to *CHIRBASE*. One can combine structure queries sample, CSPs, solvent using operators AND, OR, NOT with data queries in one search (Figure 1).

ISIS™/Base enables one to construct generic query structures which consist of combinations of partial structures to find families of the target molecule. The generic structure (Figure 2) comprises an invariant specific skeleton to which other substituents (R-groups) may be attached. R-groups may correspond to a list of specific or generic radicals and may be attached to a variable position.

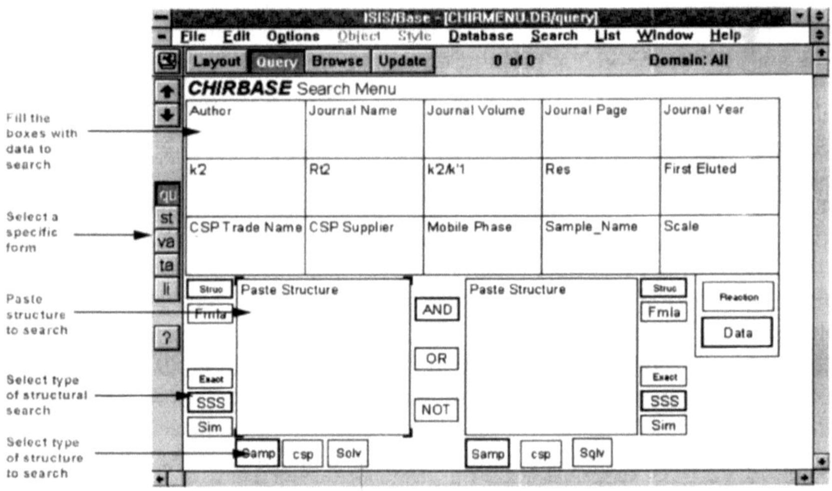

Figure 1. CHIRBASE query form in ISIS/Base (PC-client)

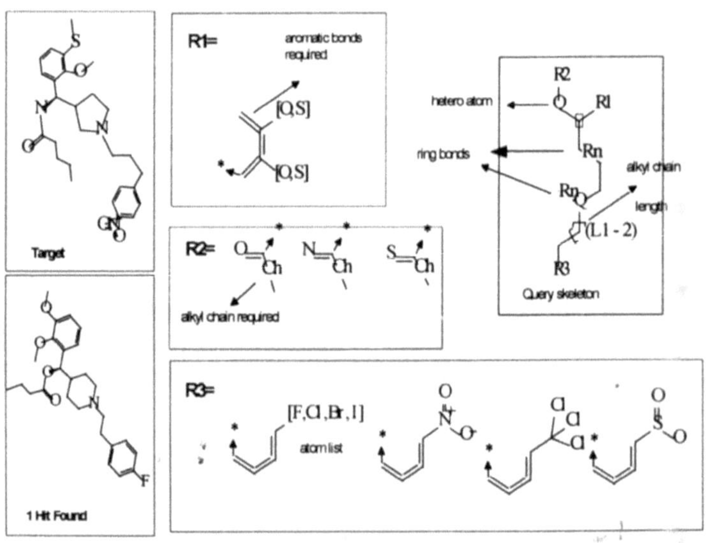

Figure 2. Example of a typical generic structural query

The other process that can predict a good candidate CSP is similarity searching. This approach can lead to an automatic classification of hits according to some measure of structural similarities estimated by an algorithm (Figure 3). An appropriate use of this procedure is for the ranking of CSPs. The measure of similarity could be a distance metric that can be used in statistical and pattern recognition techniques or in a compound recognition system.

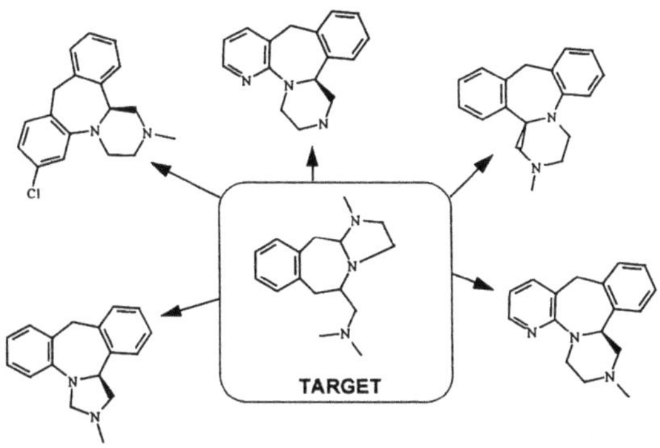

Figure 3. A 70% similarity search example

4 THE FUTURE OF CHIRBASE

While such strategies are useful, *CHIRBASE* will achieve its full potential if the prediction of a CSP for a novel molecule is based on the inspection of the 3D steric and electronic features of the CSP's binding sites to which enantiomers must bind. Such an approach will come from the availability of files of three-dimensional (3D) structures. The idea is to find in *CHIRBASE* molecules which show 3D similarity with the target compound. Then algorithms similar with those used in pharmacophore search methods could be used to correlate the 3D structures of the molecules found with the ability of those molecules to bind CSP's binding sites. Analysis of the results would allow one to isolate on the receptor (CSP) a set of active functional groups and their distances. In this way, it would be possible to analyse those key features with a variety of parameters describing molecular shape and properties by statistical techniques of multidimensional scaling to produce graphical maps of structures so that active sites can be detected visually. This information can further be used to design new synthetic CSPs.

CONCLUSION

We must emphasize here that the full exploitation of *CHIRBASE* approach requires use of a 3D context of the database. The database technique being developed links

3D searching techniques and molecular modeling. Searching facilities have been developed very rapidly and we can imagine that new computing facilities will be more and more available in the near future. Examples include the recent introduction of conformational flexibility in 3D-search in ISIS. The system now assumes during a search that a structure is not completely rigid. This kind of search which describes the conformational space of a given compound improves the efficiency of the search algorithm and permits finding more relevant hits.

CHIRBASE now contains more features than are currently searchable leaving possibilities for future enhancements. It seems reasonable to expect that such developments will lead to further advances in the area of chiral chromatographic separations.

KEYWORDS

Chiral, liquid chromatography, chiral stationary phase, database, enantiomer, resolution, REACCS, ISIS

REFERENCES

[1] Siret L., Bargmann-Leyder N., Tambuté A., Caude, M., Analusis, 20, 427-435 (1992).
[2] Francotte E. J., Chromatogr., 666, 565-601 (1994).
[3] Pirkle W.H., Pochapsky T.C., Chem. Rev., 89, 347-362 (1989).
[4] Allenmark S.G., Chromatographic Enantioseparation. Methods and Application, Ellis Horwood Ltd, Chichester (1988).
[5] Krstulovic A.M. (ed), Chiral Separations by HPLC: Applications to Pharmaceutical Compounds, Ellis Horwood Ltd, Chichester (1989).
[6] Ahuja S. (ed), Chiral Separations by Liquid Chromatography, ACS Symposium series 471, American Chemical Society, Washington D.C. (1991).
[7] Lipkowitz K.B., Demeter D.A., Zegarra R., Larter R.M., Darden T., J. Comput. Chem., 9,63-78 (1988).
[8] Lipkowitz K.B., Baker B., Anal. Chem., 62, 770-774 (1990).
[9] Seiter C., Cohan P., Int. Lab., 17 (7), 62-67 (1987).
[10] Del Rey D., Warr W.A. (ed), Graphics for chemical structures: integration with text and data, ACS Symposium Series 341, American Chemical Society, Washington, 48-61 (1987).
[11] Roussel C., Piras, P., Pure and Applied Chemistry, 65, 235-244 (1993).
[12] Koppenhoefer B., Nothdurft A., Pierrot-Sanders J., Piras P., Popescu C., Roussel C., Stiebler M., Trettin U., Chirality, 5, 213-219 (1993).

Global Recognition of the Chromatographic Profiles of Complex Matrices

Philippe GAVINI[1], *Roland FELLOUS*[1], *Eric ANGELINI*[2] *and Laure DZIUBA*[2]

[1]*Laboratoire de Chimie des Arômes, Université de Nice-Sophia Antipolis, Parc Valrose, 06108 Nice Cedex, France*
[2]*Florasynth France, 45 Bd Marcel Pagnol, 06332 Grasse Cedex, France*

Abstract

Chromatographic profiles transformed into stick profiles allow us to access the recognition and the global identification of Complex Mixtures within constituted Complex Compositions of several mixtures. We have conceived a database of automatic research of Complex Mixtures .

Resume

La transformation de profils chromatographiques en profils en bâtonnets, facilite la reconnaissance et l'identification de mélanges complexes au sein de compositions complexes constituées de plusieurs mélanges. L'élaboration d'une base de données de mélanges complexes rend automatique cette reconnaissance.

1 Introduction

The first step in determining the nature of the elements making up a complex matrix is the GC/MS analysis of the matrix. When data banks of mass spectra and indices are available, a list of the various components of the matrix hypothesized can be established. These matrices or compositions result from the combination of well-defined substances and/or mixtures resulting in turn from the association of several substances.

During GC/MS analysis only specialists thoroughly versed in mixtures can achieve global determination of the constituting elements by looking at the list of the constituents.

Numerous industrial activities (hydrocarbons, oil, paint, flavors, perfumes, pharmaceutical products, etc.) are based on the formulation of mixture association.

The problem lies in the global and automatic control of mixtures and the ability to recognize the mixtures making up compositions. This has always been a problem for formulators and it has induced us to develop a recognition technique.

The area of application selected in this work is essentially concerned with perfume and flavor. To start with, the recognition of mixtures (essential oils, extracts, bases, etc.) by the automatic comparison of their chromatographic profiles proved necessary.

This automatic profile comparison can be carried out by turning chromatograms into stick profiles, to make them look like mass spectra. Each stick is characterized by its retention index (abscissa) and its relative intensity (ordinate). Besides, each stick that characterizes a substance (whether identified or not) has its own mass spectrum.

This technique makes it possible to build up a mixture profile library[1] .

It is not necessary to have authentic invariable reference mixtures, or mixtures of which all the constituents would be identified. That way, a library can be made of any type of essences whatever their origins, whether de-terpened or not, of flavors, perfume bases or extracts, etc.

If every mixture happens to have a precise name, the library suggestions will be all the richer and will provide improved help.

In other words, a BASE X including 50 constituents, for instance, some or all of which are unknown, will be recognized as such in a composition.

Starting from this base we have evolved a recognition system of complex mixtures within a complex matrix.

Here is the method used to solve the problem. The work was performed on a Macintosh in which a library was created using 4D as software.

2 Algorithm for Complex Mixture Recognition

It consists of three stages, each including two different levels.

A detailed illustration of Stage 1 is given in Figure 1

Stage I-1 consists in selecting the complex composition to be studied from the PCC file which permits the creation of a TCC table into which all the indices and mass spectra of each constituent of the composition will be transferred.

During Stage I-2 index Ix of each peak is provided with a lower (IINF) and upper (ISUP) boundary between which the first investigation in the Complex Mixtures library will be made (MC File). Basically, it is assumed that the presence of a mixture in a composition necessarily implies the presence of the most intense component in

---

[1] Gavini P., Angelini E., Fellous R., *Parfums Cosmétiques et Arômes*, 101, 70, 1991.

the mixture. In other words, should that constituent be absent from the composition, the mixture is quite unlikely to be involved in the composition.

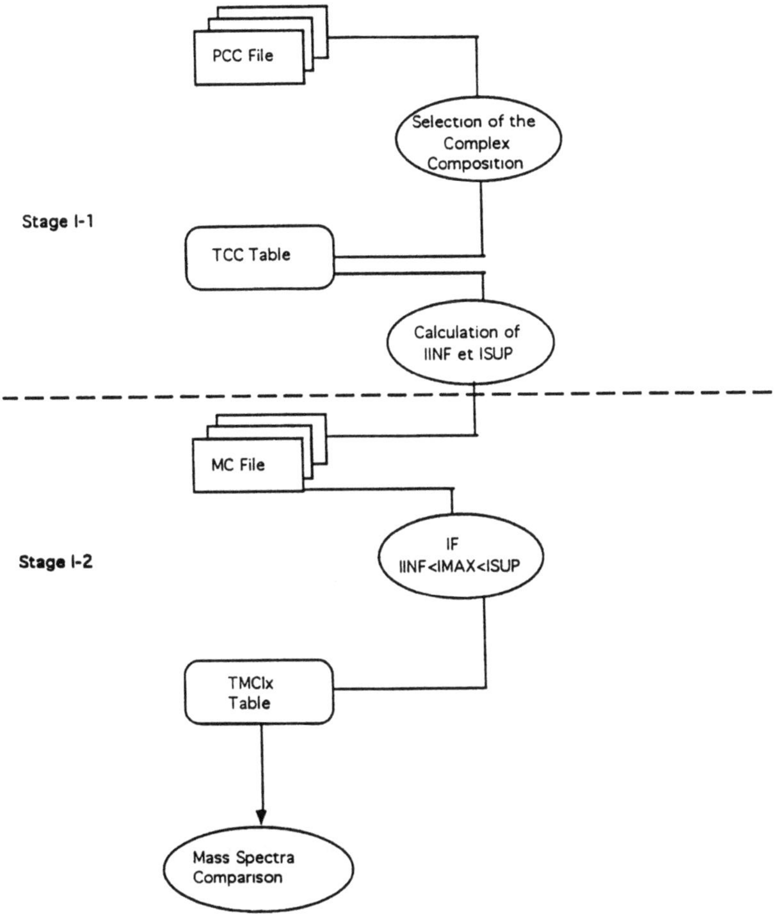

Figure 1. Detailed graph of Stage I

Thus, the first exploration is based on the notion of Imax which stands for the retention index of the highest intensity peak for the mixture. This led us to select all the Complex Mixtures whose Imax corresponds to one of the peaks in the Complex Composition. That way, the TMCIX Table can be obtained. This table shows the ID-MC identification number as well as the ID-MC identification number of its components for each complex retained.

A detailed graph of the second part of the algorithm is given in Figure 2. In stage II-1 each mass spectrum in the complex composition is compared to each Imax of each mixture previously selected. Thus, thanks to the mass spectra, the second stage enables one to verify the pre-selection made from the indices. This results in the creation of the TMCS table which collects the ID-MC identification number of each mixture finally retained in the last comparison.

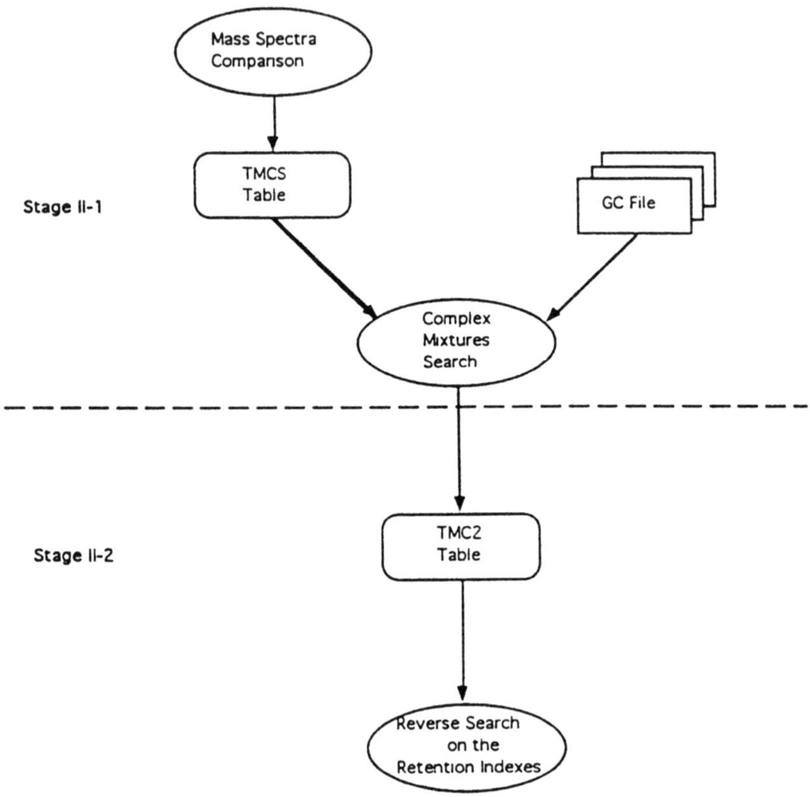

Figure 2. Detailed graph of Stage II

Stage II-2 notes all the information of the mixtures retained for further research, namely the indices, intensity and identification numbers of their components. All this is noted in table TMC2 which will be used in the third and last part of the algorithm, detailed in Figure 3.

In stage III-1, again a lower boundary and an upper boundary are determined for the retention indices of each peak in table TMC2. Then a reverse search is carried out, that is the peaks common to a mixture and a composition are detected, hence the creation of table TPCC which includes the identification number of the mass spectra retained for each mixture.

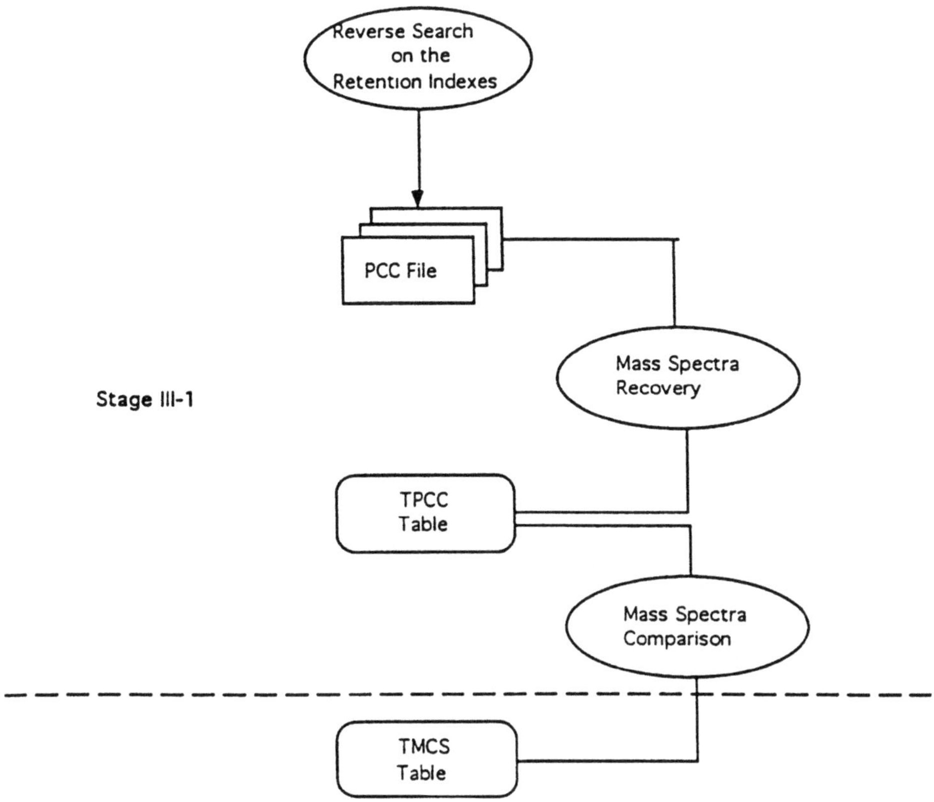

Figure 3. Detailed graph of Stage III

Last in stage III-2, the comparison procedure by mass spectra is once more applied to yield the table of results TMCS which shows the final hypotheses, selected after the various operations, together with their similarity index (S.I.). This index is calculated as follows :

$$S.I. = \frac{\sum_{i=1}^{n} (SURFCC_i \; x \; SURFMC_i)}{\sqrt{\sum_{i=1}^{n} (SURFCC_i)^2 \, x \, \sum_{i=1}^{n} (SURFCC_i)^2}}$$

where $SURFCC_i$ and $SURFMC_i$ stand respectively for the i component surface in the Complex Composition and in the Complex Mixture. The variable n stands for the number of constituents in the composition.

Looking closely at this algorithm, we note that two types of investigation are carried out : first, the so-called forward search allowing a pre-selection of the Complex Mixtures; then the so-called reverse search which consists in finding the fingerprint of the mixtures considered as sub-sets of the compositions in the Complex Composition.

3 APPLICATIONS

After creating a library of over 170 essential oils, we tested this recognition method on various compositions starting with binary, then tertiary compositions and going on to increasingly complex ones.

As an experiment designed to test our procedure we present the analysis of a series of four compositions of four essential oils, that will be named "Accords" from now on. The four Accords have a common base made up of three essences, the fourth one being different for each of them as shown in Table 1 where the figures give the constitutive percent of each essence.

NAME	ACCORD A1	ACCORD A2	ACCORD A3	ACCORD A4
ITALIAN BERGAMOT OIL	35	35	35	35
ITALIAN LEMON OIL	25	25	25	25
LAVANDIN ABRIALIS OIL	25	25	25	25
EGYPTIAN ANISE OIL	15			
STAR ANISE OIL		15		
TARRAGON OIL			15	
FENNEL OIL				15

Table 1. Constitution of the studied compositions

The four Accords analyzed by GC/MS and investigated after profile treatment enable us to reach the results in Table 2.

The results show that for the three essences common to the various Accords, the number of peaks identified remains unchanged and the similarity index varies insignificantly.

NAME	ACCORD A1		ACCORD A2		ACCORD A3		ACCORD A4	
	S.I.	Peaks	S.I.	Peaks	S.I.	Peaks	S.I.	Peaks
ITALIAN BERGAMOT OIL	86,4	16/48	88,3	16/50	87,6	16/47	88,3	16/48
ITALIAN LEMON OIL	76,4	16/48	78,3	16/50	77,9	16/47	78,5	16/48
LAVANDIN ABRIALIS OIL	50	3/48	50,9	3/50	50,9	3/47	51,1	3/48
EGYPTIAN ANISE OIL	42,1	7/48						
STAR ANISE OIL			35,9	17/50				
TARRAGON OIL					37,2	5/47		
FENNEL OIL							35,4	10/48

Table 2. Results of the four Accords analyses

Thus in Accord A1 exhibiting 48 peaks in standard injection techniques, Bergamot oil is proposed with a similarity index of 86.4%, with a recognition of 16 peaks. The constituents in a very low concentration cannot be detected due to the essence dilution in the composition.

On the other hand, we can see a significant difference in the results -whether for the index or the number of peaks retained- between Italian Lemon and Lavandin Abrialis oils. In fact, although the concentrations of both these essential oils are the same in the composition, 16 peaks are identified for the former and only three for the latter. This can be accounted for by the presence of Italian Bergamot oil which shares many more constituents with Lemon oil. These will add up in the concentration and will become more important than those from the Lavandin oil. In other words, despite equal starting concentrations, the Lavandin peaks appear to be more diluted than the Lemon ones.

This, however, does not account for the peaks found for the fourth essence since it was not so plentiful to start with. Thus, in Accord A1, seven peaks are identified for Egyptian Anise oil which includes 47 peaks in its profile while only three can be found for Lavandin oil which exhibits 49 when tested by itself according to our procedure.

Once more we can notice the peculiar behavior of the various Essential Oils namely the ones that have a very high intensity peak (superior to 80%) like Anise, Star Anise, Fennel or Tarragon oils. Besides in the examples given, the results show that the other peaks of these four essences are typical enough to be recognized in spite of their weak concentration.

The essences which, like Lavandin, exhibit several peaks of average intensity but whose nature is not characteristic enough are therefore more difficult to detect.

Nonetheless, the similarity index which increases with the concentration, reveals that the composition contains more Lavandin than Egyptian Anise. This concentration/similarity index relation can be found in a like manner in all the Accords studied.

Table 3 gives the values for the four Complex Mixtures exhibiting an anise olfactory touch varying with the Accord studied.

NAME	Accord A1		Accord A2		Accord A3		Accord A4	
	S.I.	Peaks	S.I.	Peaks	S.I.	Peaks	S.I.	Peaks
ANISE OIL	42,1	7	35,9	7			35,0	7
STAR ANISE OIL	41,9	11	35,9	17			34,8	13
TARRAGON OIL	1,68	4	1,5	4	37,2	5	1,49	4
FENNEL OIL	40,8	1	34,7	1			35,4	10

Table 3. A synthesis of the results obtained from the TMCS tables for the four Accords.

It is clear that no ambiguity can be found for Accord A3 for only Tarragon oil appears in the hypotheses of the final TMCS table. This is clear if we refer to Figure 4 showing the chromatographic profiles of the Accords.

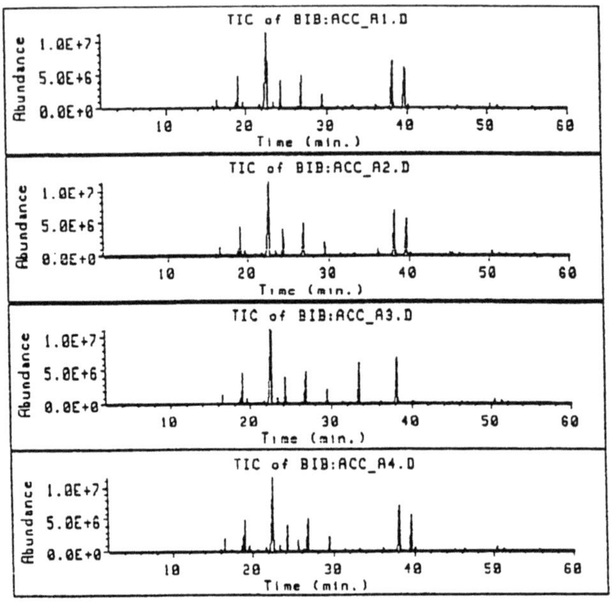

Figure 4. Superposed chromatographic profiles of the Accords.

From Figure 4 it can be seen that the chromatogram of Accord A3 is quite different from the rest. This difference is revealed by the presence of an intensity peak in the 32-to-34-minute- time zone, and by the absence of a peak in the 38-to-40-minute-zone. This implies that chromatographic distinction can be achieved by means of our recognition technique since only this essence can be found in the TMCS table.

The other tests confirm the singularity of this essence and allow us to cancel it from the list of valuable hypotheses.

As concerns Accord A2, doubt may arise if we refer to the similarity indices, but the number of peaks induces us to decide in favor of Star Anise oil, which is right.

As regards Accord A4, Fennel oil exhibits the highest similarity index which is confirmed by close analysis of the complete TMCS Table. Comparison of the main peak contribution is not significant since it is the same for all three Essential Oils. The second intensity peak, however, allows us to know we are dealing with Fennel oil. In fact, in the case of Fennel, the peak which represents 14.82% of the Mixture can be found as 5.61% in the composition and no other essence retained in the final table shows any of it. As a result only Fennel oil provides this constituent. This can be seen on the first line of the following table (Table 4). The second and third line show the values for the second intensity peak of Star Anise oil and Anise oil respectively.

Accord A4	**Fennel oil**	**Star Anise oil**	**Anise oil**
5,61	14,82		
1,30	2,46	6,08	0,18
1,25	0,32	0,60	4,93

Table 4. An extract of the final TMCS table for Accord A4. Contribution of the second intensity peak.

We can see that the Star Anise oil second peak which makes up 1.3% of Accord A4 can also be provided in sufficient amounts by Fennel oil and not by Anise oil. On the other hand, Anise oil whose concentration is 1.25 % in the composition proves to be in insufficient amounts in Fennel and Star Anise oils. It can be found, however, in Bergamot and Lemon oils, which accounts for its intensity if we consider that Fennel oil is involved in this analysis.

Last, in the case of Accord 1, as for Fennel oil in Accord A4, the similarity index is highest for Anise oil, and the examination of the peaks confirms this hypothesis.

The results obtained show that the similarity index is always linked to the concentration and reveals the relative importance of mixtures. Whenever, in a given number of hypotheses, the number of peaks is identical, it permits choosing the most likely one. Similarly, whenever the similarity index is the same, the number of peaks recognized will reveal the suitable hypothesis. Should any doubt remain, it can be cleared by a close analysis of the final TMCS table.

CONCLUSION

The technique developed for the global recognition of mixtures involved in a complex composition may make monitoring easier or give easier access to information on the nature of the mixtures making up a Complex Matrix.

This involves a practical procedure, and to start with the creation of a library of stick profiles for the mixtures. It includes the retention index, the relative intensity and the mass spectrum for each constituent.

Such a library can be used to check whether raw materials are genuine or not.

Starting with the analysis of a composition, the first step consists in selecting the mixtures whose most intense constituent corresponds to one of the substances in the composition. After automatic checking by means of mass spectra and selection roundup, a reverse search is carried out. This last step seeks the fingerprint of a mixture in the composition analyzed. It is not necessary to identify all the constituents at each step to find out the mixtures that make up a complex matrix.

While this approach developed does not call upon specific knowledge of essential oils, it makes it possible to solve most of the problems encountered when dealing with mixtures. However, any expert knowledge in that area would prove beneficial and increase recognition efficiency.

One of the most interesting aspects of the method used is that it can be easily extended to other fields and other analytical couplings.

KEY WORDS

Retention Index, mass spectrum, PBM, standard exchange format of spectroscopic data , JCAMP , complex mixtures , complex compositions .

MORPHOLOGY RECOGNITION BY MEANS OF CHIRAL GAS CHROMATOGRAPHY

Roland FELLOUS, Louisette LIZZANI and Michel-André LOISEAU

Laboratoire de Chimie des Arômes -Université de Nice-Sophia Antipolis, Parc Valrose, 06108 Nice Cédex 2, France

ABSTRACT

Morphology recognition of enantiomeric structures are differentiated by chiral stationary phase. Effects of substituent and behavior of various families of racemics, esters of secondary alchohols, γ and ε lactones, are examined. In a homologous series it becomes possible to predict retention data .

RESUME

La reconnaissance morphologique de composés énantiomètres est utilisée à l'aide de phases stationnaires chirales. Le rôle des substituants alkyles est abordé à l'aide du comportement chromatographique de différentes familles (esters d'alcools secondaires, γ et ε lactones). Dans une série homologue il devient possible de prévoir avec succès les indices de rétention .

1 SUMMARY

Enantiomeric structures interact with chiral stationary GC phases, inducing a distinct retention time for each spatial structure.

Such separations are valuable to authentify chiral substances of biological origin, such as, for instance, those entering into flavors and fragrances.

Modified cyclodextrins as chiral stationary phases give access to the separation of enantiomers through the building up of diastereotopic complexes whose stabilities are characterized by specific interactions originating in inclusion of phenomena that are equivalent to solute/solvent interactions.

The separation of racemic ε-lactones is described. Some examples illustrate the analogy in the comportment of terms within a homologous series.

2 INTRODUCTION

The resolution of racemics has always been a matter of preoccupation for chemists and biochemists. Biochemists early came across the problem raised by the specific interactions linked with the chirality of the molecule used, especially in the therapeutic field.

Numerous cases can be found in this field. Let us mention for instance Adrenaline, where enantiomer *R* (-) is 50 times more active than enantiomer *S* (+), or vitamine C, where only L(+) ascorbic acid exhibits the antiscorbutic property. In other compounds, one enantiomer can be active while another develops highly undesirable properties, including toxic ones. Thalidomide belongs to this category.

In the same way, such specific biological interactions can be found within our olfacto-gustatory system. Thus, *S* (+) Aspargine is bitter when *R* (-) Asparagine is sweet; *S* (+) Carvone from Carvi's oil has a taste and a smell totally different from its reverse *R* (-) from *Mentha Viridis* Linnaeus. This fast expanding field of study often leads to amazing results which, however, provide information enabling scientists to understand the mechanisms and interactions in the phenomena of taste and smell.

This is why simple precise and rapid methods for the evaluation of excess enantiomerics are extremely useful to establish (re)activity/structure relationships.

As far back as 1988 Juvancz and Szejtli [1] showed that CDs modified by etherification and/or esterification, are stable at high temperatures, and used in GC as stationary phases, could resolve racemics. This has led to the development of chiral phases often suited to the resolution of particular racemics, or to specific problems involving both the size of the cavity of CDs and the nature of the functional transformation.

Keim and Coll.[2] have drawn up a balance list of the resolutions for 150 compounds according to the structure of the CD used.

The mechanisms of enantiodifferentiation of CDs, as chirals phases for capillary GC, are not yet totally clear. It seems that optical antipodes form evolutive diastereotopic complexes within the phase. The different retention times point to a difference in free energy of the two complexes and illustrate an enantioselective behavior.

To determine the interaction mechanisms of "solute racemic gas/solvent chiral liquid" we have examined the chromatographic behavior of a series of secondary alcohols acetate, of 2-alkyl cyclohexanones, 6-alkyl ε-lactones and 4-alkyl γ-lactones versus permethylated β-CD and of γ-CD (Lipodex E). Those compounds share structural characteristics, namely one polar site and one nonpolar chain.

Figure 1 shows the enantiodifferentiation obtained for a series of acetates in the case of the permethylated phase of β-CD (phase I). All racemic esters are resolved.

Enantiomer *S* is always eluted first. The order of elution was determined by the co-injection of one pure enantiomer.

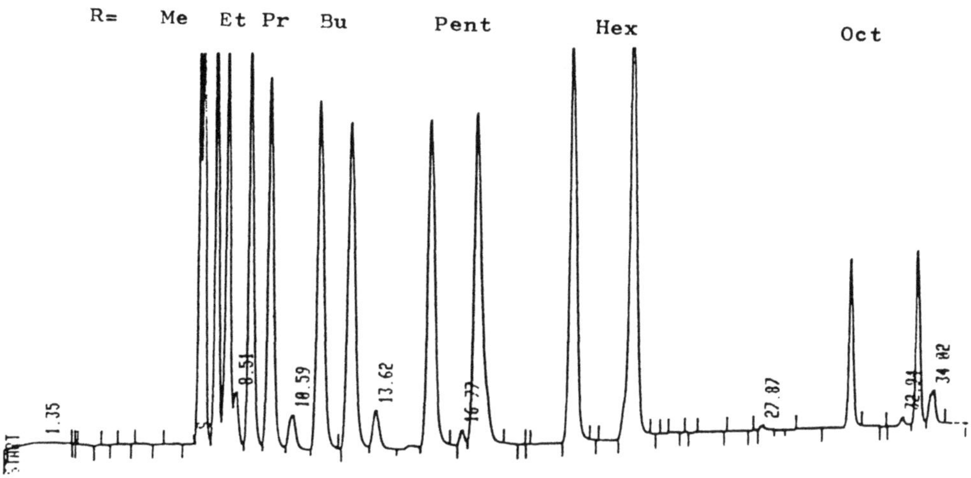

Only the first term (methyl) of 2-alkyls cyclohexanones, is not separated and resolution improves with the length of the alkyl chain, as is the case with acetates. All the series of racemics of γ-lactones is resolved. The retention order of the enantiomers is based on the comparison with commercial references of pure optical substances.

The kinetic resolution of racemics of ε-lactones by enzymatic hydrolyse, using Pig-Liver Esterase [3], has enabled us to obtain each of the enantiomers enriched to more than 98%.

In the series of four families studied enantiomer *S* is always eluted first.

Table 1 shows, for the four families, the values of retention indices of enantiomers R and S , calculated from Van Der Dool' s [4] 'formula , with the n-alcanes of C_6 to C_{28} used as references. The phase **I** used, synthesized at the laboratory, [5], consists of permethylated β-CD. Phase **II** (Lipodex E Macherey Nagel) is a commercial γ-CD. Only the series of ε-lactones is resolved by means of two chiral phases.

To study the effects of substituents and the behavior analogies of the four families, we have used the concepts proposed by J.-E. Dubois et al.[6] which permit classifying all the relations according to different types of correlations.

In the same chiral phase the relation between the retention indices of isomer *R* as compared to those isomer *S* shows that both substituents have a similar influence on the phenomenon observed.

Table 1.

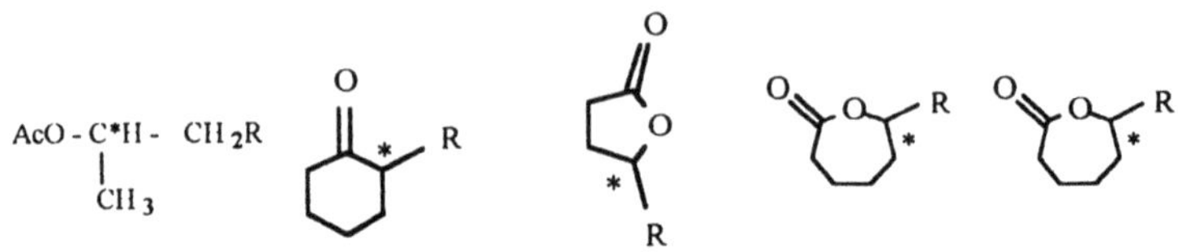

Phase	I		I		II		I		II	
Alkyl=	Indices S	Indices R	Indices S	Indices R	Indices S	Indices R	Indices S	Indices R	Indices S	Indices R
Me	818	832			1499	1506	1503	1520	1480	1509
Et	881	910	1225	1229	1562	1576	1547	1562	1509	1592
nPr	951	986	1286	1290	1576	1597	1626	1638	1689	1711
nBu	1040	1070	1380	1383	1609	1628	1738	1748	1709	1729
nPent	1136	1171	1496	1502	1676	1688	1865	1877	1812	1824
nHex	1241	1284	1627	1633	1762	1771	1993	2006	1899	1907
nHept			1757	1762	1854	1863	2119	2130	1997	2003
nOct	1461	1515	1875	1880	1948	1956	2237	2247	2102	2107
nNon										
nDodeca			2311	2315						

Whatever the population examined a linear relation is obtained (Figure 2).

Figure 2.

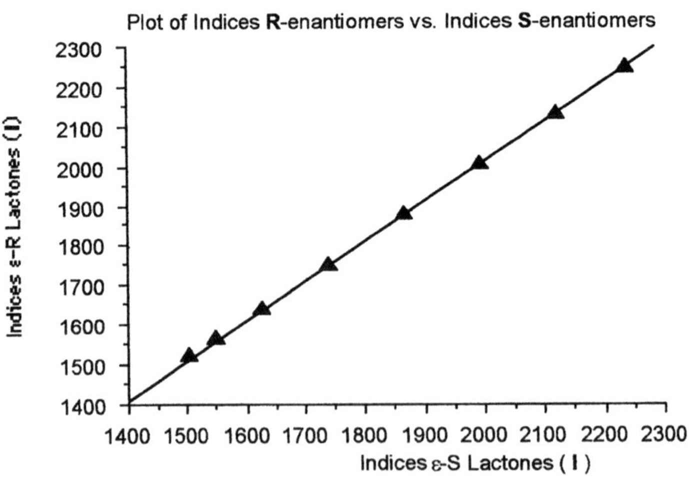

Table 2 shows the parameter of these linear relations.

Table 2. Variations of slope " a " in equation : I(R) = a I(S) for different racemics

	a	r	p
acetates	1,033	1	7
alkyls-cyclohexanones	1,003	1	8
alkyls γ-lactones	1,007	1	8
alkyls ε-lactones (**II**)	1,012	1	8
alkyls e-lactones (**I**)	1,007	1	8

The results in Table 2 show the analogy of interaction of the two enantiomeric series for the same chiral phase

As the slope "a" of the different straight lines is close to one it implies that the alkyl variable, nonpolar group acts in a like manner in the two enantiomers series of all four families . At a first approximation, the coefficient "a", which is in relation with the resolution factor of the column, is characteristic of the the specific interaction of the *R* (or *S*) series on this chiral phase.

The relation I(R) = a. I (S) shows that substituents behave homogeneously in the case of physico-chemical interactions with the chiral phase. The slight difference in energy of the diastereotopics complexes in a series is due to the minimal variation of the polar effects of alkyl groups. The enantiodifferentiation of a racemic solute results from complex and manifold interactions in which the surface and inclusion phenomena play an essential role. Those interactions are thought to have dipole-dipole, Van Der Waals or Keesom as the main factor.

This has allowed us to establish linear relations of free enthalpy of the following type:

$$I = \rho^* . \sigma^*$$

where I stands as the retention indices and σ^* is a constant for the polar effects of alkyl group, ρ^* represents the sensitivity of the phenomenon to polar effects.

In those relations and in all cases the methyl group stands out and is excluded from the correlation.

The values of the σ^* retained are adapted to the chromatographic phenomenon and take mass variation into account [7]. In a homologous series with temperature linear programming retention times grow linearly with the number "n" of carbon atoms.

If the relation

$$I\ (R\ \text{or}\ S) = m.(nC) + q$$

is verified, it implies that the interactions in a series do not alter the elution orders of the optical antipodes, and that the enthalpic and entropic values remain continuous without changing signs in the experimental temperature scale used.

Figure 3 represents the application of this relation in the case of ε-lactones and Table 6 shows the parameters "m" and "q" of the relation.

Figure 3.

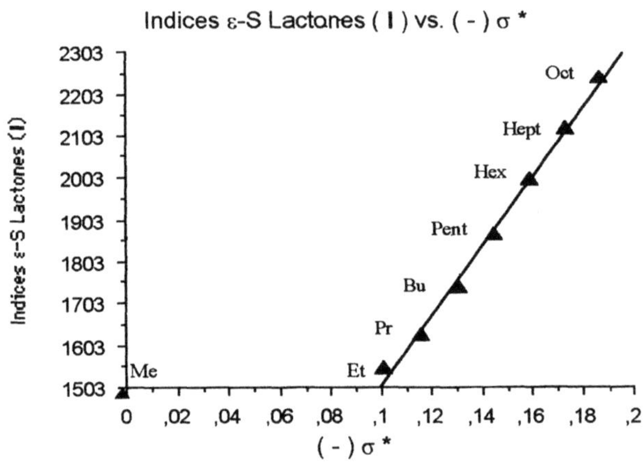

Table 6.

Families	m	q	r	p
alkylcyclohexanones (S)	113,1	959	0,988	8
alkylcyclohexanones (R)	113,6	963	0,998	8
alkyls γ-lactones (S)	62,2	1406	0,975	8
alkyls γ-lactones (R)	61,5	1421	0,978	8
acetates (S)	92,3	600	0,995	7
acetates (R)	97,2	609	0,995	7
alkyls ε-lactones (S) **(II)**	86,2	1394	0,995	8
alkyls ε-lactones (R) **(II)**	82,4	1427	0,995	8
alkyls ε-lactones (S) **(I)**	109,8	1334	0,993	8
alkyls ε-lactones (R) **(I)**	109,0	1350	0,992	8

Those parameters depend on the nature of the skeleton of the family studied, as well as on the properties of the chiral phase used.

The value of "m" reflects the interactions with the chiral phase for a given series. The value of the intercept "q" stands , within the limits of experimental errors, for the retention indice of the first term in the series (R = H / nC = 0). It can be seen from this table that the calculated value of the retention indice of the γ-lactone is of the order of 1406-1421, within the limit of experimental errors, when the measured value is I = 1418.

This relation fails for δ-lactones when δ-6 lactone is eluted after δ-7 and δ-8 [8]. If a temperature of coalescence for two optical antipodes [9] exists:

- Δ(ΔG°) in Gibbs-Helmholtz's equation becomes equal to 0, and the coalescence temperature can be expressed as follows :
  T_{coal} = Δ(ΔH°) / Δ(ΔS°). Elution order of enantiomers is reversed if the temperature is higher than T_{coal}.

CONCLUSION

The phenomena from which enantiodifferentiation originate are complex. While in liquid chromatography such phenomena seem to involve hydrogen bonds and inclusion phenomena, it seems unlikely in gas chromatography which makes it possible to resolve chiral hydocarbures. The forming of hydrogen bonds sometimes appears to be an unfavourable factor towards separation. Determining thermodynamic parameters will make it possible to understand chiral recognition better. Such recognition involves hydrophobic interactions within the CD cavity as well as dipolar interactions with the CD outer surface. This resolution technique has already proved useful in the understanding of a few plant biosyntheses and the development of new biotechnological methods.

MATERIAL AND METHODS

* ε-*lactone synthesis* and *enzymatic hydrolysis* are described in [3].

* gas phase chromatography : The chromatograph used is a HP 5890 model fitted with a Flame Ionisation Detector (FID) and linked to an integrator-recorder HP 3392 A.

A Lipodex E (Macherey Nagel) capillary column was used (fused silica, L = 25 m.: ϕ = 0.25 mm.). The other column is a permethylated β-CD (fused silica, L = 50 m.: ϕ = 0.25 mm) synthesized in our laboratory [5].

The carrier gas is nitrogen ; split 1/100, t° injector 230°C, t° detector 250°C. Column inlet pressure is 60 kPa for Lipodex E and 100 kPa for permethylated β-CD ; The test conditions range from 60°C to 180°C with a programmed temperature of 2°/mn for Lipodex E and range from 130°C to 180°C with a programmed temperature of 2°/mn for β-CD. The samples are in a 10% solution in dichloromethane.

KEYWORDS

Chiral gas chromatography, modified cyclodextrins, data retention indexes, racemics esters, γ and ε lactones, linear relationships .

REFERENCES

[1] Alexander G., Juvancz Z., Szejtli J., J.HRC & CC,11,110 (1988).
[2] Keim W., Kohnes A., Meltzow W., Romer H., J.HRC & CC,14,507 (1991).
[3] Fellous R., Lizzani-Cuvelier L., Loiseau M.A., Tetrahedron Assymetry,5,343 (1994).
[4] Van Den Dool H., Kratz P.D., J.chromatogr.,11,463 (1963).
[5] Fellous R., Lizzani-Cuvelier L., Loiseau M.A., J.HRC & CC.,13,785 (1990).
[6] Dubois J.E., Aranda A., Compt.Rend.Acad.Sci. ,269,1451 (1969).
[7] Fellous R., Luft R., Rabine J.P., J.Chromatogr.,140,137 (1977).
[8] Dietrich A., Maas B., Karl V., Kreis P., Lehmann D., Weber B., Mosandl A., J.HRC & CC.,15,176 (1992).
[9] Berthod A., Li W., Armstrong W., Anal. Chem. , 64, 873 (1992).

Quantum Chemistry Techniques Applied to the Modelling of the Enantioselective Hydrogenation Mechanism of α-Ketoesters

Olivier SCHWALM[1], *Jacques WEBER*[1], *Brune MINDER*[2] *and Alfont BAIKER*[2]

[1] *Department of Physical Chemistry, University of Geneva, 30 quai Ernest-Ansermet, CH-1211 Geneva 4, Switzerland*
[2] *Department of Chemical Engineering and Industrial Chemistry, Swiss Federal Institute of Technology, ETH Zentrum, CH-8092 Zürich, Switzerland*

Abstract

Enantio-differentiation in the asymmetric hydrogenation of α-ketoesters to α-hydroxyesters over platinum catalysts modified with cinchona-alkaloid modifiers occurs through interaction of the ketoester with the cinchona modifier. The properties of the reactant and the structure of the probable transition complex methyl pyruvate (substrate) - cinchonidine (modifier) are calculated using quantum chemistry techniques.

The calculations suggest that protonated cinchonidine interacts with the substrate and that the crucial interaction occurs via hydrogen bonding of the protonated quinuclidine nitrogen and the oxygen of the α-carbonyl moiety of methyl pyruvate. In this complex the methyl pyruvate is stabilized in a half-hydrogenated state which may be the key for the enantio-differentiation.

Resume

L'hydrogénation énantiosélective des α-cétoesters en α-hydroxyesters a lieu sur une surface de Pt/Al_2O_3 modifiée par la cinchonidine qui interagit avec le cétoester. Les propriétés du réactif et la structure d'un probable complexe de transition méthyle pyruvate (substrat) - cinchonidine (modificateur) sont calculées en utilisant des techniques de chimie quantique.

Les résultats des calculs suggèrent que la cinchonidine protonée interagit avec le substrat et que l'interaction cruciale fait intervenir une liaison hydrogène entre l'azote protoné de la quinuclidine et l'oxygène du carbonyle α du méthyle pyruvate. Ce complexe ressemble à un état semi-hydrogéné du substrat et c'est un intermédiaire possible pour l'hydrogénation énantiosélective des α-cétoesters.

1 INTRODUCTION

Enantioselective hydrogenation using solid catalysts has been extensively studied in the cases of β-ketoesters interacting with tartrate modified nickel [1,2] and of α-ketoesters with cinchona modified platinum catalysts (scheme 1) [3]. The mechanisms of these reactions are still poorly understood and theoretical investigations should help to rationalize the interactions of the chiral modifier with the substrate, solvent and metal surface.

In a recent experimental study, Wehrli et al. have suggested that a very specific interaction takes place between substrate (P), modifier (C) and platinum surface [4]. In particular, the heteroaromatic system (quinoline) of the modifier is probably responsible for the adsorption on the platinum surface, whereas the quinuclidine part is involved in the interaction with the substrate leading to asymmetric induction. On this particular point, Blaser et al. have concluded that the nitrogen atom of quinuclidine plays the major role in substrate complexation, the OH group of this fragment probably not being involved. A possible mechanism of interaction has been then inferred, according to which "the α-ketoester interacts preferentially with the modified sites on the platinum surface where the enantioselective reaction takes place" [5].

The purpose of the present study is to rationalize the interaction between the chiral cinchonidine modifier and the methyl pyruvate substrate, and to explore whether the theoretical results are in agreement with earlier experimental findings [6] and proposals [7]. For this purpose quantum chemistry techniques at both ab initio and semiempirical levels are used.

2 THEORY

The geometries of the model systems P-NH_3 and P-NH_4^+ were fully optimized using the 3-21G basis set at the SCF and MP2 levels and with constraints, i.e. with a pyruvate system frozen at its optimum geometry, using the 6-31++G basis set at the SCF level, as implemented in the Gaussian 90 series of programs [8] running on a Silicon Graphics workstation. To evaluate the propensity of P and C to behave as electrophiles or as nucleophiles, the P-H^- and P-H^+ model complexes were investigated using a semiempirical model. The latter is based on a local reactivity index consisting of the interaction energy $E_{int}(\mathbf{r})$ between a substrate and an incoming electrophilic or nucleophilic model reactant located in **r** calculated using extended-Hückel (EH) wavefunctions.[9, 10] Then, ab initio calculations on the P-C protonated complex were performed at the SCF level using the 3-21G basis set as implemented in the HONDO 8.5 series of programs [11] running on an IBM SP2 parallel machine.

3 RESULTS AND DISCUSSION

Figure 1 depicts the ab initio equilibrium geometries of the P-NH_3 and P-NH_4^+ complexes together with EH isoenergy surfaces calculated for a nucleophilic and electrophilic attack of P, respectively. Note that the zone of highest reactivity for nucleophilic attack is located between the ester- and keto-carbonyl groups of P, whereas the most favorable sites for electrophilic attack lie approximately in the regions of the lone pairs of the oxygen atoms. Actually, docking of P to a protonated form of C via a hydrogen bond interaction is energetically more favorable, as shown by a much more stabilizing complexation energy for the P-NH_4^+ case (by 25 kcal/mol) predicted by the ab initio calculations. This result indicates that such a complex is thermodynamically more stable and therefore likely to predominate under conditions where cinchonidine can be protonated. We suggest that the high enantioselectivity of methyl pyruvate hydrogenation obtained in acetic acid [12] is mainly due to the enhanced affinity of the reactant for a protonated form of C existing in this medium.

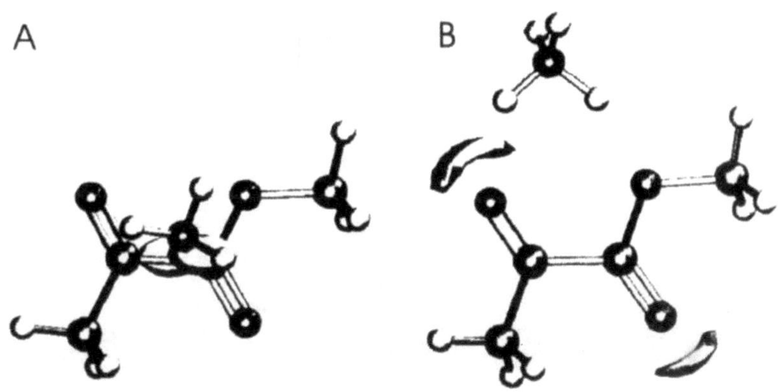

Figure 1. Ab initio equilibrium geometries of the P-NH_3 and P-$NH4^+$ complexes together with the EH isoenergy surfaces of E_{int} calculated for the nucleophilic attack on P (-6 kcal/mol, black) (A) and for electrophilic attack (-150 kcal/mol) (B).

As a protonated modifier molecule is more likely to react with P than an unprotonated one, we have investigated the possible conformations of the complex formed upon interaction of protonated C with P. Two stable forms have been fully optimized using the ab initio technique, and Fig. 3 displays the one which exhibits the most favorable steric interactions with the platinum catalyst. Again the crucial interaction takes place via hydrogen bonding between the quinuclidine nitrogen and the oxygen atom on the α-carbonyl. It is known that there is a favorable interaction (π–bonding) for the quinoline system of C when laying parallel to the metal surface. Using the molecular graphics package Molekel [14], it is possible to dock the whole complex onto a Pt(111) surface [15], considering only steric interactions, in such a way as to have the quinoline aromatic ligand parallel to the surface. It is then found that the P substrate is located near the surface, i.e. in a position which is favorable for a transfer of an adsorbed hydrogen atom. The P-C complex adsorbed on the platinum surface is thus stabilized by hydrogen bonding transforming the substrate P in a half-hydrogenated state, which suggests that this process is likely to be at the origin of the reaction mechanism of enantioselective hydrogenation of α-ketoesters.

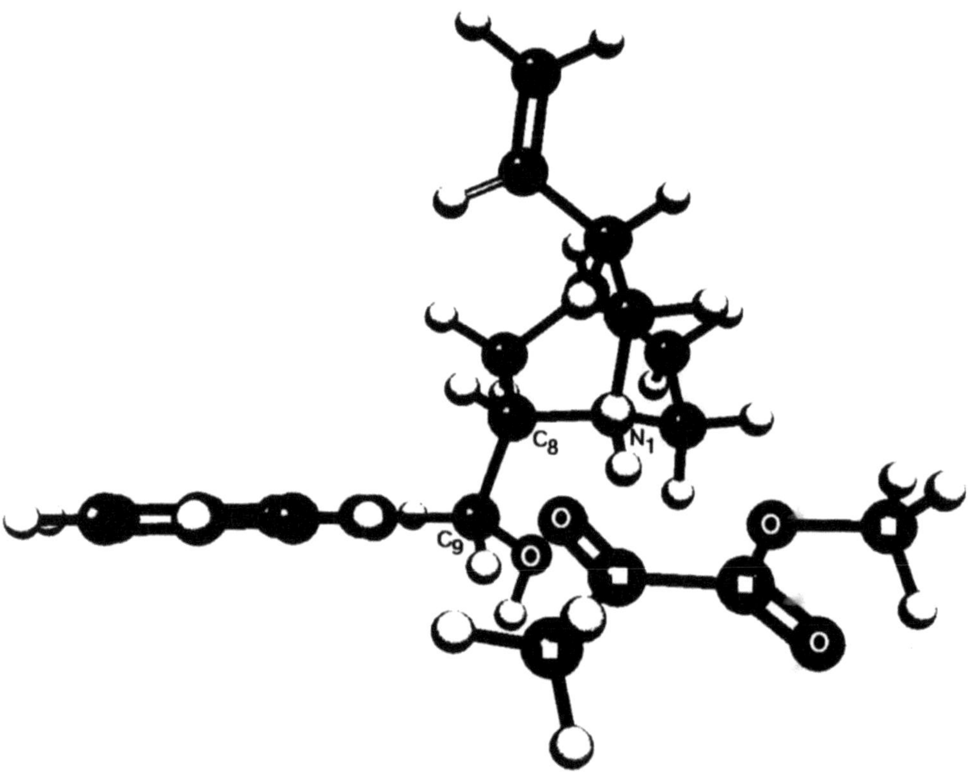

Figure 2. Protonated C-P complex fully optimized according to 3-21 G basis set (E_{int}= -21.0 kcal/mol).

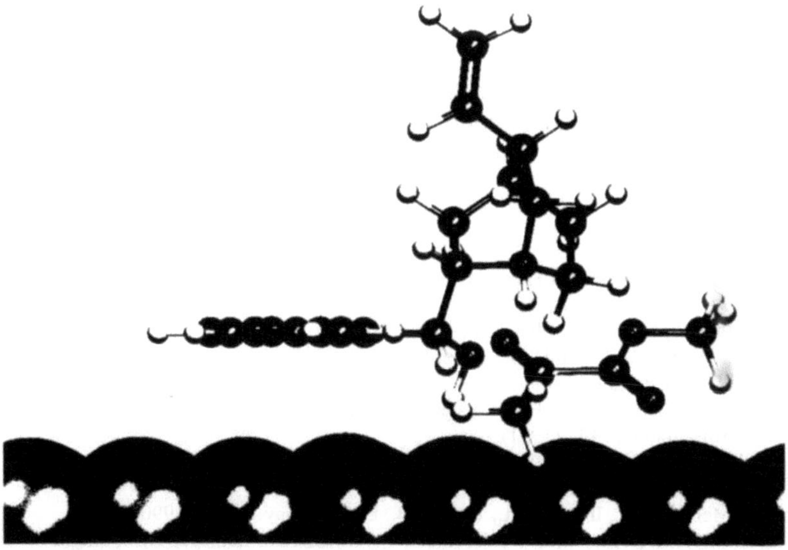

Figure 3. Side view of the optimized structure of the protonated C-P complex accomodated on a platinum (111) surface.

The molecular modelling approach, taking into account P, C and the steric constraints imposed by the adsorption on the platinum surface, leads to a reasonable explanation for the enantio-differentiation of this system.

CONCLUSION

Our theoretical studies aiming at finding the structure of the complex formed upon interaction of P with C indicate that C protonated at the quinuclidine nitrogen forms a more stable complex with P than unprotonated C. In the P-C protonated complex the substrate is transformed into a half-hydrogenated state by hydrogen bonding between the protonated quinuclidine nitrogen of C and the oxygen at the α-carbonyl of P. However, our theoretical prediction of the P-C complex was made for an ideal case, since solvent effects and a quantum description of the interaction with the platinum surface atoms are not considered. Furthermore, the question of whether the P-C complex is formed in solution or by coadsorption of P and C on the platinum surface is still open.

ACKNOWLEDGEMENT

Financial support by the Swiss National Science Foundation is kindly acknowledged.

KEYWORDS

Enantioselective hydrogenation, α-ketoesters, α-hydroxyesters, cinchonidine, modifier, Pt/alumina, molecular modelling

REFERENCES

[1] Bartok M., Stereochemistry of Heterogeneous Metal Catalysts, Wiley, New York, p.511 (1985).
[2] Izumi Y., Adv. Catal., 32, 215 (1983).
[3] Orito Y., Imai S., Niwa S., J. Chem. Soc., Jpn. 670 (1980).
[4] Wehrli J.T., Baiker A., Monti D.M., Blaser H.U., Jalett H.P., J. Mol. Catal., 57 245 (1989).
[5] Blaser, H.U., Jalett H.P., Monti D.M., Baiker A., Wehrli J.T., in: Structure-Activity and Selectivity Relationships in Heterogeneous Catalysis, Grasselli R.K., Sleight A.W. (eds), Elsevier, Amsterdam, p. 147 (1991).
[6] Blaser H.U., Jalett H.P., Monti D.M., Baiker A., Wehrli J.T., Stud. Surf. Sci. Catal. 67 147 (1991).
[7] Bond G., Meheux P.A., Ibbotson A., Wells P.B., Catal. Today 10 371 (1990).
[8] Frisch M.J., Head-Gordon M., Trucks G.W., Foresman J.B., Schlegel H.B., Raghavachari K., Robb M.A., Binkley J.S., Gonzales C., Defrees D.J., Fox D.J., Whiteside R.A., Seeger R., Melius C.F., Baker J., Martin R.L., Kahn L.R., Stewart J.J.P., Topiol S., Pople J., Gaussian Inc., Pittsburgh PA (1990).
[9] Daul C., Goursot A., Morgantini P.Y., Weber J., Int. J. Quant. Chem. 38 623 (1990).
[10] Weber J., Morgantini P.Y., Eisenstein O., J. Mol. Struct. (Theochem) 254 343 (1992).
[11] Dupuis M., Johnston F., Marquez A., HONDO 8.5 from CHEM-Station, IBM Corporation, Neighborhood Road, Kingston, N.Y. 12401 (1994).
[12] Blaser H.U., Jalett H.P., Wiehl J., J. Mol. Catal. 68 215 (1991).
[13] Soriaga M.P., Binamira-Soriaga E., Hubbard A.T., Benziger J.B., Pang K.W.P., Inorg. Chem. 24, 65 (1985)
[14] Fluekiger P.F., Ph. D. Thesis N° 2561, University of Geneva (1992).
[15] Schwalm O., Minder B., Weber J., Baiker A., Catal. Lett. 23 271 (1994).

INDEX

Springer-Verlag and the Environment

We at Springer-Verlag firmly believe that an international science publisher has a special obligation to the environment, and our corporate policies consistently reflect this conviction.

We also expect our business partners – paper mills, printers, packaging manufacturers, etc. – to commit themselves to using environmentally friendly materials and production processes.

The paper in this book is made from low- or no-chlorine pulp and is acid free, in conformance with international standards for paper permanency.

MIX
Papier aus verantwortungsvollen Quellen
Paper from responsible sources
FSC® C105338

If you have any concerns about our products,
you can contact us on
ProductSafety@springernature.com

In case Publisher is established outside the EU,
the EU authorized representative is:
Springer Nature Customer Service Center GmbH
Europaplatz 3, 69115 Heidelberg, Germany

Printed by Libri Plureos GmbH
in Hamburg, Germany